Fault Tree Handbook

U.S. Nuclear Regulatory
Commission

Fault Tree Handbook

Date Published: January 1981

W. E. Vesely, U.S. Nuclear Regulatory Commission
F. F. Goldberg, U.S. Nuclear Regulatory Commission

N. H. Roberts, University of Washington
D. F. Haasl, Institute of System Sciences, Inc.

Systems and Reliability Research
Office of Nuclear Regulatory Research
U.S. Nuclear Regulatory Commission
Washington, D.C. 20555

For sale by the U.S. Government Printing Office
Superintendent of Documents, Mail Stop: SSOP, Washington, DC 20402-9328

TABLE OF CONTENTS

Introduction vii

I. Basic Concepts of System Analysis I-1

    1. The Purpose of System Analysis I-1
    2. Definition of a System I-3
    3. Analytical Approaches I-7
    4. Perils and Pitfalls I-9

II. Overview of Inductive Methods II-1

    1. Introduction II-1
    2. The "Parts Count" Approach II-1
    3. Failure Mode and Effect Analysis (FMEA) II-2
    4. Failure Mode Effect and Criticality Analysis (FMECA) II-4
    5. Preliminary Hazard Analysis (PHA) II-4
    6. Fault Hazard Analysis (FHA) II-5
    7. Double Failure Matrix (DFM) II-5
    8. Success Path Models II-10
    9. Conclusions II-12

III. Fault Tree Analysis—Basic Concepts III-1

    1. Orientation III-1
    2. Failure vs. Success Models III-1
    3. The Undesired Event Concept III-3
    4. Summary III-4

IV. The Basic Elements of a Fault Tree IV-1

    1. The Fault Tree Model IV-1
    2. Symbology—The Building Blocks of the Fault Tree IV-1

V. Fault Tree Construction Fundamentals V-1

    1. Faults vs. Failures V-1
    2. Fault Occurrence vs. Fault Existence V-1
    3. Passive vs. Active Components V-2
    4. Component Fault Categories: Primary, Secondary, and Command . . V-3
    5. Failure Mechanism, Failure Mode, and Failure Effect V-3
    6. The "Immediate Cause" Concept V-6
    7. Basic Rules for Fault Tree Construction V-8

VI. Probability Theory—The Mathematical Description of Events VI-1

    1. Introduction   VI-1
    2. Random Experiments and Outcomes of Random Experiments. VI-1
    3. The Relative Frequency Definition of Probability VI-3
    4. Algebraic Operations with Probabilities VI-3
    5. Combinatorial Analysis VI-8
    6. Set Theory: Application to the Mathematical Treatment
        of Events VI-11
    7. Symbolism VI-16
    8. Additional Set Concepts VI-17
    9. Bayes' Theorem VI-19

VII. Boolean Algebra and Application to Fault Tree Analysis VII-1

    1. Rules of Boolean Algebra VII-1
    2. Application to Fault Tree Analysis VII-4
    3. Shannon's Method for Expressing Boolean Functions in
        Standardized Forms VII-12
    4. Determining the Minimal Cut Sets or Minimal Path Sets of a
        Fault Tree VII-15

VIII. The Pressure Tank Example VIII-1

    1. System Definition and Fault Tree Construction VIII-1
    2. Fault Tree Evaluation (Minimal Cut Sets) VIII-12

IX. The Three Motor Example IX-1

    1. System Definition and Fault Tree Construction IX-1
    2. Fault Tree Evaluation (Minimal Cut Sets) IX-7

X. Probabilistic and Statistical Analyses X-1

    1. Introduction X-1
    2. The Binomial Distribution X-1
    3. The Cumulative Distribution Function X-7
    4. The Probability Density Function X-9
    5. Distribution Parameters and Moments X-10
    6. Limiting Forms of the Binomial: Normal, Poisson X-15
    7. Application of the Poisson Distribution to System Failures—
        The So-Called Exponential Distribution X-19
    8. The Failure Rate Function X-22
    9. An Application Involving the Time-to-Failure Distribution X-25
    10. Statistical Estimation X-26
    11. Random Samples X-27
    12. Sampling Distributions X-27
    13. Point Estimates—General X-28

TABLE OF CONTENTS

	14.	Point Estimates—Maximum Likelihood	X-30
	15.	Interval Estimators	X-35
	16.	Bayesian Analyses	X-39
XI.		Fault Tree Evaluation Techniques	XI-1
	1.	Introduction	XI-1
	2.	Qualitative Evaluations	XI-2
	3.	Quantitative Evaluations	XI-7
XII.		Fault Tree Evaluation Computer Codes	XII-1
	1.	Overview of Available Codes	XII-1
	2.	Computer Codes for Qualitative Analyses of Fault Trees	XII-2
	3.	Computer Codes for Quantitative Analyses of Fault Trees	XII-6
	4.	Direct Evaluation Codes	XII-8
	5.	PL-MOD: A Dual Purpose Code	XII-11
	6.	Common Cause Failure Analysis Codes	XII-12
		Bibliography	BIB-1

INTRODUCTION

Since 1975, a short course entitled "System Safety and Reliability Analysis" has been presented to over 200 NRC personnel and contractors. The course has been taught jointly by David F. Haasl, Institute of System Sciences, Professor Norman H. Roberts, University of Washington, and 'members of the Probabilistic Analysis Staff, NRC, as part of a risk assessment training program sponsored by the Probabilistic Analysis Staff.

This handbook has been developed not only to serve as text for the System Safety and Reliability Course, but also to make available to others a set of otherwise undocumented material on fault tree construction and evaluation. The publication of this handbook is in accordance with the recommendations of the Risk Assessment Review Group Report (NUREG/CR-0400) in which it was stated that the fault/event tree methodology both can and should be used more widely by the NRC. It is hoped that this document will help to codify and systematize the fault tree approach to systems analysis.

CHAPTER I — BASIC CONCEPTS OF SYSTEM ANALYSIS

1. The Purpose of System Analysis

The principal concern of this book is the fault tree technique, which is a systematic method for acquiring information about a system.* The information so gained can be used in making decisions, and therefore, before we even define system analysis, we will undertake a brief examination of the decisionmaking process. Decisionmaking is a very complex process, and we will highlight only certain aspects which help to put a system analysis in proper context.

Presumably, any decision that we do make is based on our present knowledge about the situation at hand. This knowledge comes partly from our direct experience with the relevant situation or from related experience with similar situations. Our knowledge may be increased by appropriate tests and proper analyses of the results—that is, by experimentation. To some extent our knowledge may be based on conjecture and this will be conditioned by our degree of optimism or pessimism. For example, we may be convinced that "all is for the best in this best of all possible worlds." Or, conversely, we may believe in Murphy's Law: "If anything can go wrong, it will go wrong." Thus, knowledge may be obtained in several ways, but in the vast majority of cases, it will not be possible to acquire all the relevant information, so that it is almost never possible to eliminate all elements of uncertainty.

It is possible to postulate an imaginary world in which no decisions are made until all the relevant information is assembled. This is a far cry from the everyday world in which decisions are forced on us by time, and not by the degree of completeness of our knowledge. We all have deadlines to meet. Furthermore, because it is generally impossible to have all the relevant data at the time the decision must be made, we simply cannot know all the consequences of electing to take a particular course of action. Figure I-1 provides a schematic representation of these considerations.

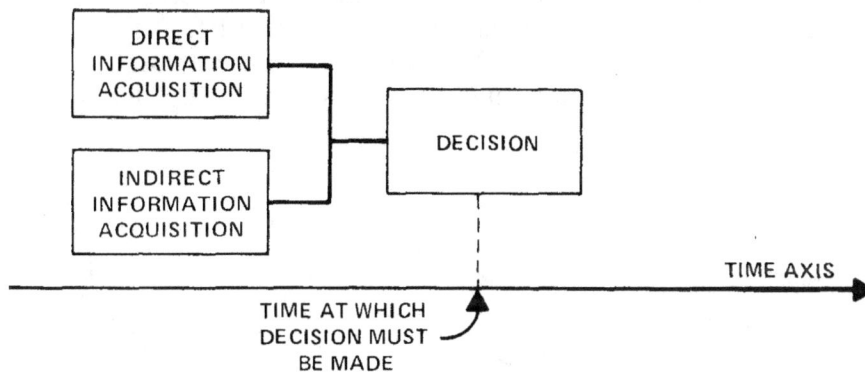

Figure I-1. Schematic Representation of the Decisionmaking Process

---

*There are other methods for performing this function. Some of these are discussed briefly in Chapter II.

The existence of the time constraint on the decisionmaking process leads us to make a distinction between good decisions and correct decisions. We can classify a decision as good or bad whenever we have the advantage of retrospect. I make a decision to buy 1000 shares of XYZ Corporation. Six months later, I find that the stock has risen 20 points. My original decision can now be classified as good. If, however, the stock has plummeted 20 points in the interim, I would have to conclude that my original decision was bad. Nevertheless, that original decision could very well have been correct if all the information available at the time had indicated a rosy future for XYZ Corporation.

We are concerned here with making correct decisions. To do this we require:

(1) The identification of that information (or those data) that would be pertinent to the anticipated decision.

(2) A systematic program for the acquisition of this pertinent information.

(3) A rational assessment or analysis of the data so acquired.

There are perhaps as many definitions of system analysis as there are people working and writing in the field. The authors of this book, after long thought, some controversy, and considerable experience, have chosen the following definition:

System analysis is a directed process for the orderly and timely acquisition and investigation of specific system information pertinent to a given decision.

According to this definition, the primary function of the system analysis is the acquisition of information and not the generation of a system model. Our emphasis (at least initially) will be on the process (i.e., the acquisition of information) and not on the product (i.e., the system model). This emphasis is necessary because, in the absence of a directed, manageable, and disciplined process, the corresponding system model will not usually be a very fruitful one.

We must decide what information is relevant to a given decision before the data gathering activity starts. What information is essential? What information is desirable? This may appear perfectly obvious, but it is astonishing on how many occasions this rationale is not followed. The sort of thing that may happen is illustrated in Figure I-2.

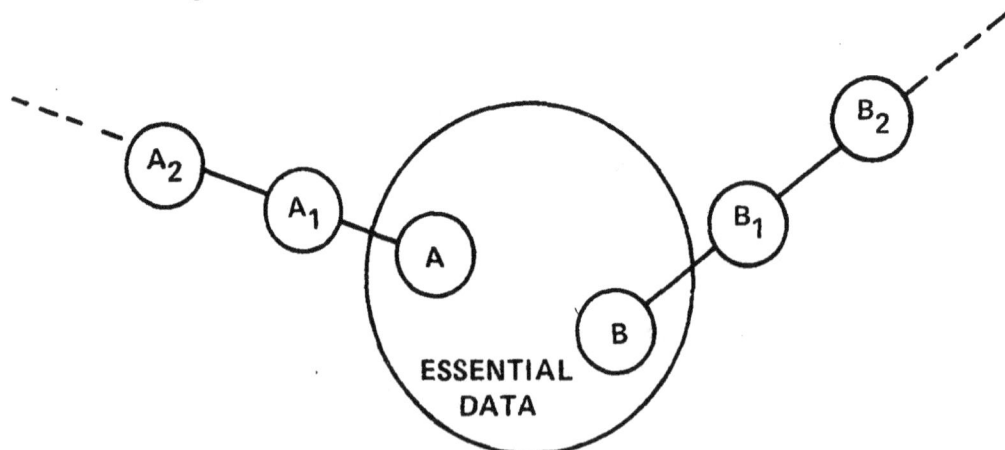

Figure I-2. Data Gathering Gone Awry

The large circle represents the information that will be essential for a correct decision to be made at some future time. Professor Jones, who is well funded and who is a "Very Senior Person," is an expert in sub-area A. He commences to investigate this area, and his investigation leads him to some fascinating unanswered questions indicated in the sketch by A_1. Investigation of A_1 leads to A_2, and so on. Notice, however, that Professor Jones' efforts are causing him to depart more and more from the area of essential data. Laboratory Alpha is in an excellent position to study sub-area B. These investigations lead to B_1 and B_2 and so on, and the same thing is happening. When the time for decision arrives, all the essential information is not available despite the fact that the efforts expended would have been able to provide the necessary data if they had been properly directed.

The nature of the decisionmaking process is shown in Figure I-3. Block A represents certified reality. Now actual reality is pretty much of a "closed book," but by experimentation and investigation (observations of Nature) we may slowly construct a perception of reality. This is our system model shown as block B. Next, this model is analyzed to produce conclusions (block C) on which our decision will be based. So our decision is a direct outcome of our model and if our model is grossly in error, so also will be our decision. Clearly, then, in this process, the greatest emphasis should be placed on assuring that the system model provides as accurate a representation of reality as possible.

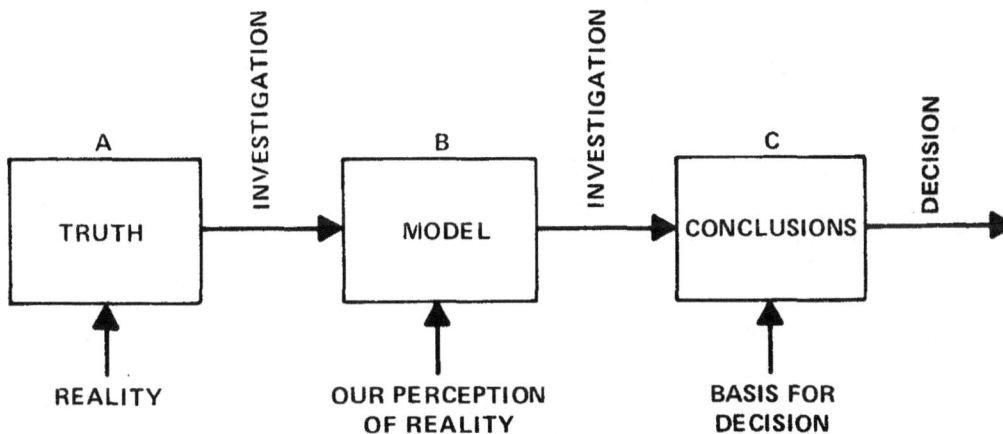

Figure I-3. Relationship Between Reality, System
Model, and Decision Process

2.   Definition of a System

We have given a definition of the process of system analysis. Our next task is to devise a suitable definition for the word "system." In common parlance we speak of "the solar system," "a system of government," or a "communication system," and in so doing we imply some sort of organization existing among various elements that mutually interact in ways that may or may not be well defined. It seems reasonable, then, to establish the following definition:

A system is a <u>deterministic entity</u> comprising an <u>interacting collection</u> of <u>discrete elements</u>.

From a practical standpoint, this is not very useful and, in particular cases, we must specify what aspects of system performance are of immediate concern. A system performs certain functions and the selection of particular performance aspects will dictate what kind of analysis is to be conducted. For instance: are we interested in whether the system accomplishes some task successfully; are we interested in whether the system fails in some hazardous way; or are we interested in whether the system will prove more costly than originally anticipated? It could well be that the correct system analyses in these three cases will be based on different system defintions.

The word "deterministic" in the definition implies that the system in question be identifiable. It is completely futile to attempt an analysis of something that cannot be clearly identified. The poet Dante treated the Inferno as a system and divided it up into a number of harrowing levels but, from a practical standpoint, such a system is not susceptible to identification as would be, for example, the plumbing system in my home. Furthermore, a system must have some purpose—it must do. something. Transportation systems, circulating hot water piping systems, local school systems all have definite purposes and do not exist simply as figments of the imagination.

The discrete elements of the definition must also, of course, be identifiable; for instance, the individual submarines in the Navy's Pacific Ocean Submarine Flotilla. Note that the discrete elements themselves may be regarded as systems. Thus, a submarine consists of a propulsion system, a navigation system, a hull system, a piping system, and so forth; each of these, in turn, may be further broken down into subsystems and sub-subsystems, etc.

Note also from the definition that a system is made up of parts or subsystems that interact. This interaction, which may be very complex indeed, generally insures that a system is not simply equal to the sum of its parts, a point that will be continually emphasized throughout this book. Furthermore, if the physical nature of any part changes—for example by failure—the system itself also changes. This is an important point because, should design changes be made as a result of a system analysis, the new system so resulting will have to be subjected to an analysis of its own. Consider, for example, a four-engine aircraft. Suppose one engine fails. We now have a new system quite different from the original one. For one thing, the landing characteristics have changed rather drastically. Suppose two engines fail. We now have six different possible systems depending on which two engines are out of commission.

Perhaps the most vital decision that must be made in system definition is how to put external boundaries on the system. Consider the telephone sitting on the desk. Is it sufficient to define the system simply as the instrument itself (earpiece, cord and cradle), or should the line running to the jack in the wall be included? Should the jack itself be included? What about the external lines to the telephone pole? What about the junction box on the pole? What about the vast complex of lines, switching equipment, etc., that comprise the telephone system in the local area, the nation, the world? Clearly some external boundary to the system must be established and this decision will have to be made partially on the basis of what aspect of system performance is of concern. If the immediate problem is that the bell is not loud enough to attract my attention when I am in a remote part of the house, the external system boundary will be fairly closed in. If the problem involves static on the line, the external boundary will be much further out.

It is also important in system definition to establish a limit of resolution. In the telephone example, do I wish to extend my analysis to the individual components (screws, transmitter, etc.) of which the instrument is composed? Is it necessary to descend to the molecular level, the atomic level, the nuclear level? Here again, a decision can be made partially on the basis of the system aspect of interest.

What we have said so far can be represented as in Figure I-4. The dotted line separates the system from the environment in which it is embedded. Thus, this dotted line constitutes an external boundary. It is sometimes useful to divide the system into a number of subsystems, A, B, C, etc. There may be several motivations for doing this; most of them will be discussed in due course. Observe also that one of the subsystems, F, has been broken down, for purposes of the analysis, into its smallest sub-subsystems. This constitutes a choice of an internal boundary for the system. The smallest sub-subsystems, a, b, c, etc., are the "discrete elements" mentioned in the general definition of a system.

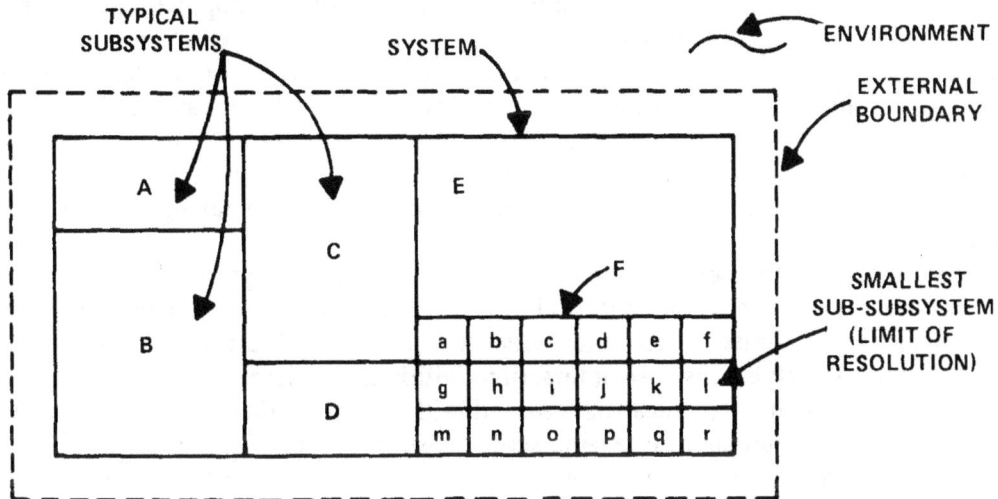

Figure I-4. System Definition: External and Internal Boundaries

The choice of the appropriate system boundaries in particular cases is a matter of vital importance, for the choice of external boundaries determines the comprehensiveness of the analysis, whereas the choice of a limit of resolution limits the detail of the analysis. A few further facets of this problem will be discussed briefly now, and will be emphasized throughout the book, especially in the applications.

The system boundaries we have discussed so far have been physical boundaries. It is also possible, and indeed necessary in many cases, to set up temporal or time-like boundaries on a system. Consider a man who adopts the policy of habitually trading his present car in for a new one every two years. In this example, the system is the car and the system aspect of interest is the maintenance policy. It is clear that, under the restriction of a two-year temporal boundary, the maintenance policy adopted will be one thing; it will be quite a different thing if the man intends to run the car for as long as possible. In some applications, the system's physical boundaries might actually be functions of time. An example of this would be a system whose temporal boundaries denote different operating phases or different design modifications. After each phase change or design modification, the physical boundaries are subject to review and possible alteration.

The system analyst must also ask the question, "Are the chosen system boundaries feasible and are they valid in view of the goal of the analysis?" To reach certain conclusions about a system, it may be desirable to include a large system "volume" within the external boundaries. This may call for an extensive, time-consuming analysis. If the money, time and staff available are inadequate for this task and more efficient analysis approaches are not possible, then the external boundaries must be "moved in" and the amount of information expected to result from the analysis must be reduced. If I am concerned about my TV reception, it might be desirable to include the state of the ionosphere in my analysis, but this would surely be infeasible. I would be better advised to reduce the goals of my analysis to a determination of the optimum orientation of my roof antenna.

The limit of resolution (internal boundary) can also be established from considerations of feasibility (fortunately!) and from the goal of the analysis. It is possible to conduct a worthwhile study of the reliability of a population of TV sets without being concerned about what is going on at the microscopic and submicroscopic levels. If the system failure probability is to be calculated, then the limit of resolution should cover component failures for which data are obtainable. At any rate, once the limit of resolution has been chosen (and thus the "discrete elements" defined), it is interactions at this level with which we are concerned; we assume no knowledge and are not concerned about interactions taking place at lower levels.

We now see that the external boundaries serve to delineate system outputs (effects of the system on its environment) and system inputs (effects of the environment on the system); the limits of resolution serve to define the "discrete elements" of the system and to establish the basic interactions within the system.

The reader with a technical background will recognize that our definition of a system and its boundaries is analogous to a similar process involved in classical thermodynamics, in which an actual physical boundary or an imaginary one is used to segregate a definite quantity of matter ("control mass") or a definite volume ("control volume"). The inputs and outputs of the system are then identified by the amounts of energy or mass passing into or out of the bounded region. A system that does not exchange mass with its environment is termed a "closed system" and a closed system that does not exchange energy with its surroundings is termed an "adiabatic system" or an "isolated system." A student who has struggled with thermodynamic problems, particularly flow problems, will have been impressed with the importance of establishing appropriate system boundaries before attempting a solution of the problem.

A good deal of thought must be expended in the proper assignment of system boundaries and limits of resolution. Optimally speaking, the system boundaries and limits of resolution should be defined before the analysis begins and should be adhered to while the analysis is carried out. However, in practical situations the boundaries or limits of resolution may need to be changed because of information gained during the analysis. For example, it may be found that system schematics are not available in as detailed a form as originally conceived. The system boundaries and limits of resolution, and any modifications, must be clearly defined in any analysis, and should be a chief section in any report that is issued.

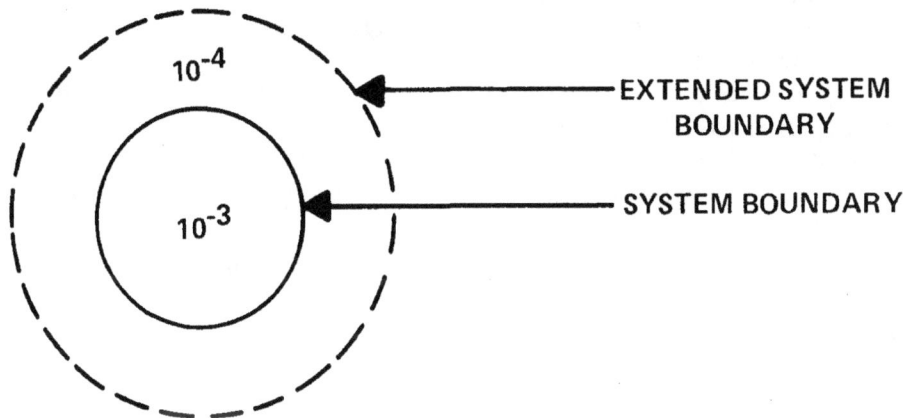

Figure I-5. Effect of System Boundaries on Event Probabilities

To illustrate another facet of the bounding and resolution problem, consider Figure I-5. The solid inner circle represents our system boundary inside of which we are considering events whose probabilities of occurrence are, say, of order 10^{-3} or greater. If the system boundaries were extended (dotted circle) we would include, in addition, events whose probabilities of occurrence were, say, of order 10^{-4} or greater. By designing two-fold redundancy into our restricted system (solid circle) we could reduce event probabilities there to the order of $(10^{-3})(10^{-3}) = 10^{-6}$ but then the probabilities of events that we are ignoring become overriding, and we are suffering under the delusion that our system is two orders of magnitude safer or more reliable than it actually is. When due consideration is not devoted to matters such as this, the naive reliability calculator will often produce such absurd numbers as 10^{-16} or 10^{-18}. The low numbers simply say that the system is not going to fail by the ways considered but instead is going to fail at a much higher probability in a way not considered.

3.   Analytical Approaches

Because we are concerned, in this volume, with certain formal processes or models, it should come as no great suprise that these processes or models can be categorized in exactly the same way as are the processes of thought employed in the human decisionmaking process. There are two generic analytical methods by means of which conclusions are reached in the human sphere: induction and deduction. It is necessary at this point to discuss the respective characteristics of these approaches.

Inductive Approaches

Induction constitutes reasoning from individual cases to a general conclusion. If, in the consideration of a certain system, we postulate a particular fault or initiating condition and attempt to ascertain the effect of that fault or condition on system operation, we are constructing an inductive system analysis. Thus, we might inquire into how the loss of some specified control surface affects the flight of an airplane or into how the elimination of some item in the budget affects the overall operation of a school district. We might also inquire how the non-insertion of given control rods affects a scram system's performance or how a given initiating event, such as a pipe rupture, affects plant safety.

Many approaches to inductive system analysis have been developed and we shall devote Chapter II to a discussion of the most important among them. Examples of this method are: Preliminary Hazards Analysis (PHA), Failure Mode and Effect Analysis (FMEA), Failure Mode Effect and Criticality Analysis (FMECA), Fault Hazard Analysis (FHA), and Event Tree Analysis.

To repeat—in an inductive approach, we assume some possible component condition or initiating event and try to determine the corresponding effect on the overall system.

Deductive Approaches

Deduction constitutes reasoning from the general to the specific. In a deductive system analysis, we postulate that the system itself has failed in a certain way, and we attempt to find out what modes of system/component* behavior contribute to this failure. In common parlance we might refer to this approach as a "Sherlock Holmesian" approach. Holmes, faced with given evidence, has the task of reconstructing the events leading up to the crime. Indeed, all successful detectives are experts in deductive analysis.

Typical of deductive analyses in real life are accident investigations: What chain of events caused the sinking of an "unsinkable" ship such as the Titanic on its maiden voyage? What failure processes, instrumental and/or human, contributed to the crash of a commercial airliner into a mountainside?

The principal subject of this book, Fault Tree Analysis, is an example of deductive system analysis. In this technique, some specific system state, which is generally a failure state, is postulated, and chains of more basic faults contributing to this undesired event are built up in a systematic way. The broad principles of Fault Tree Analysis, as well as details relating to the applications and evaluation of Fault Trees, are given in later chapters.

In summary, inductive methods are applied to determine what system states (usually failed states) are possible; deductive methods are applied to determine how a given system state (usually a failed state) can occur.

---

*A component can be a subsystem, a sub-subsystem, and sub-sub-subsystem, etc. Use of the word "component" often avoids an undesirable proliferation of "subs."

4. Perils and Pitfalls

In the study of systems there are dangerous reefs which circumscribe the course which the analyst must steer. Most of these problem areas assume the role of interfaces: subsystem interfaces and disciplinary interfaces.

Subsystem Interfaces

Generally, a system is a complex of subsystems manufactured by several different subcontractors or organizational elements. Each subcontractor or organizational element takes appropriate steps to assure the quality of his own product. The trouble is that when the subsystems are put together to form the overall system, failure modes may appear that are not at all obvious when viewed from the standpoint of the separate component parts.

It is important that the same fault definitions be used in analyses which are to be integrated, and it is important that system boundaries and limits of resolution be clearly stated so that any potential hidden faults or inconsistencies will be identified. The same event symbols should be used if the integrated system is to be evaluated or quantified. Interface problems often lie in control systems and it is best not to split any control system into "pieces." Systems which have control system interfaces (e.g., a spray system having an injection signal input) can be analyzed with appropriate "places" left for the control analysis which is furnished as one entity. These "places," or transfers, will be described later in the fault tree analysis discussions.

Disciplinary Interfaces

Difficulties frequently arise because of the differing viewpoints held by people in different disciplines or in different areas of employment. The circuit designer regards his black box as a thing of beauty and a joy forever, a brainchild of his own creation. He handles it gently and reverently. The user, on the other hand, may show no such reverence. He drops it, kicks it and swears at it with gay abandon.

One of the authors, as a mere youth, was employed as a marine draftsman. The principal shop project was drawing up plans for mine-sweepers. The draftsmen were divided into groups. There was a hull section, a wiring section, a plumbing section, etc., each section working happily within its own technical area. When an attempt was made to draw up a composite for one of the compartments (the gyro room, in this case), it was found that hull features and plumbing fixtures were frequently incompatible, that wiring and plumbing often conflicted with vent ducts, and indeed, that the gyro room door could not be properly opened because of the placement of a drain from the head on the upper deck. This exercise demonstrated the unmistakable need for system integration.

Other conflicts can readily be brought to mind: the engineering supervisor who expects quantitative results from his mathematical section, but gets only beautiful existence proofs; the safety coordinator who encumbers the system with so many safety devices that the reliability people have trouble getting the system to work at all; and so on.

As an example of an interface between operational and maintenance personnel, consider a system that is shut down for an on-line maintenance check for 5 minutes every month, and suppose that the probability of system failure due to hardware failure is 10^{-6} per month. Then, on a monthly basis, the total probability that the system will be <u>unavailable</u> is the sum of its unavailability due to hardware failures and its unavailability due to the maintenance policy or:

$$10^{-6} + \frac{1/12}{720} \cong 10^{-6} + 10^{-4}.$$

where
    10^{-6} = system unavailability due to hardware failure per month
    720   = number of hours in a month
    1/12 = hours required for maintenance check per month

Note that the probability of system unavailability due to our maintenance policy (only 5 minutes downtime per month) is greater by two orders of magnitude than the probability that the system will be down because of hardware failure. In this case the best maintenance policy may be none at all!

The system analyst (system integrator) must be unbiased enough and knowledgeable enough to recognize interface problem areas when they occur—and they will occur.

CHAPTER II — OVERVIEW OF INDUCTIVE METHODS

1. Introduction

In the last chapter we defined the two approaches to system analysis: inductive and deductive. The deductive approach is Fault Tree Analysis—the main topic of this book. This chapter is devoted to a discussion of inductive methods.

We have felt it necessary to devote a full chapter to inductive methods for two reasons. First of all, these techniques provide a useful and illuminating comparison to Fault Tree Analysis. Second, in many systems (probably the vast majority) for which the expenses and refinements of Fault Tree Analysis are not warranted, the inductive methods provide a valid and systematic way to identify and correct undesirable or hazardous conditions. For this reason, it is especially important for the fault tree analyst to be conversant with these alternative procedures.

In everyday language the inductive techniques provide answers to the generic question, "What happens if--?" More formally, the process consists of assuming a particular state of existence of a component or components and analyzing to determine the effect of that condition on the system. In safety and reliability studies the "state of existence" is a fault. This may not necessarily be true in other areas.

For systems that exhibit any degree of complexity (i.e., for most systems), attempts to identify all possible system hazards or all possible component failure modes—both singly and in combination—become simply impossible. For this reason the inductive techniques that we are going to discuss are generally circumscribed by considerations of time, money and manpower. Exhaustiveness in the analysis is a luxury that we cannot afford.

2. The "Parts Count" Approach

Probably the simplest and most conservative (i.e., pessimistic) assumption we can make about a system is that any single component failure will produce complete system failure. Under this assumption, obtaining an upper bound on the probability of system failure is especially straightforward. We simply list all the components along with their estimated probabilities of failure. The individual component probabilities are then added and this sum provides an upper bound on the probability of system failure. This process is represented below:

Component	Failure Probability
A	f_A
B	f_B
$\cdot$	$\cdot$
$\cdot$	$\cdot$
$\cdot$	$\cdot$

where F, the failure probability for the system, is equal to $f_A + f_B + \ldots$

The failure probabilities can be failure rates, unreliabilities, or unavailabilities depending on the particular application (these more specific terms will be covered later).

For a particular system, the Parts Count technique can provide a very pessimistic estimate of the system failure probability and the degree of pessimism is generally not quantifiable. The "Parts Count" technique is conservative because if critical components exist, they often appear redundantly, so that no single failure is actually catastrophic for the system. Furthermore, a component can often depart from its normal operating mode in several different ways and these failure modes will not, in general, all have an equally deleterious effect on system operation. Nevertheless, let us see what results the Parts Count approach yields for the simple parallel configuration of two amplifiers shown in Figure II-1.

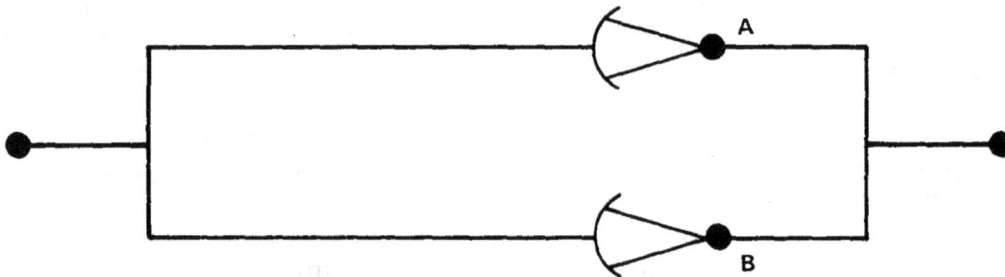

Figure II-1. A System of Two Amplifiers in Parallel

Suppose the probability of failure of amplifier A is 1×10^{-3} and the probability of failure of amplifier B is 1×10^{-3}, i.e., $f_A = 1 \times 10^{-3}$ and $f_B = 1 \times 10^{-3}$. Because the parallel configuration implies that system failure would occur only if both amplifiers fail, and assuming independence of the two amplifiers, the probability of system failure is $1 \times 10^{-3} \times 1 \times 10^{-3} = 1 \times 10^{-6}$. By the parts count method, the component probabilities are simply summed and hence the "parts count system failure probability" is $1 \times 10^{-3} + 1 \times 10^{-3} = 2 \times 10^{-3}$ which is considerably higher than 1×10^{-6}.

The parts count method thus can give results which are conservative by orders of magnitude if the system is redundant. When the system does have single failures, then the parts count method can give reasonably accurate results. Because all the components are treated as single failures (any single component failure causes system failure), any dependencies among the failures are covered, i.e., the parts count method covers multiple component failures due to a common cause.* Finally, the parts count method can also be used in sensitivity studies; if the system or subsystem failure probability does not impact or does not contribute using the parts count method, then it will not impact or contribute using more refined analyses.

3.   Failure Mode and Effect Analysis (FMEA)

Inasmuch as the Parts Count approach is very simplistic and can give very conservative results, other more detailed techniques have been devised. We first

---

*Common cause failures will be discussed in subsequent chapters.

discuss Failure Mode and Effect Analysis and return for a closer look at the system shown in Figure II-1.

We recognize that amplifiers can fail in several ways and our first task is the identification of these various failure modes. The two principal ones are "open" and "short" but suppose that our analysis has also detected 28 other modes (e.g., weak signal, intermittent ground, etc.). A short of any amplifier is one of the more critical failure modes inasmuch as it will always cause a failure of the system. We now describe a table containing the following information:

(1) Component designation

(2) Failure probability (failure rates or unavailabilities are some of the specific characteristics used)

(3) Component failure modes

(4) Percent of total failures attributable to each mode

(5) Effects on overall system, classified into various categories (the two simplest categories are "critical" and "non-critical").

The result for our redundant amplifier system might be as in Table II-1.

Table II-1. Redundant Amplifier Analysis

1	2	3	4	5	
Component	Failure Probability	Failure Mode	% Failures by Mode	Critical	Effects Non-Critical
A	1×10^{-3}	Open	90		X
		Short	5	X (5×10^{-5})	
		Other	5	X (5×10^{-5})	
B	1×10^{-3}	Open	90		X
		Short	5	X (5×10^{-5})	
		Other	5	X (5×10^{-5})	

Based on prior experience with this type of amplifier, we estimate that 90% of amplifier failures can be attributed to the "open" mode, 5% of them to the "short" mode, and the balance of 5% to the "other" modes. We know that whenever either amplifier fails shorted, the system fails so we put X's in the "Critical" column for these modes; "Critical" thus means that the single failure causes system failure. On the other hand, when either amplifier fails open, there is no effect on the system from the single failure because of the parallel configuration. What is the criticality of the other 28 failure modes? In this example we have been conservative and we are considering them all as critical, i.e., the occurrence of any one causes system failure. The numbers shown in the Critical column are obtained from multiplying the appropriate percentage in Column 4 by 10^{-3} from Column 2.

Based on the table, we can now more realistically calculate the probability of system failure from single causes, considering now only those failure modes which are critical. Adding up the critical column, Column 5, we obtain probability of system failure $= 5 \times 10^{-5} + 5 \times 10^{-5} + 5 \times 10^{-5} + 5 \times 10^{-5} = 2 \times 10^{-4}$. This is a less conservative result compared to 2×10^{-3} obtained from the parts count method where the critical failure modes were not separated. The difference between the two system results can be large, i.e., an order of magnitude or more, as in our example, if the critical failure modes are a small percentage of the total failure modes (e.g., 10% or less).

In FMEA (and its variants) we can identify, with reasonable certainty, those component failures having "non-critical" effects, but the number of possible component failure modes that can realistically be considered is limited. Conservatism dictates that unspecified failure modes and questionable effects be deemed "critical" (as in the previous example). The objectives of the analysis are to identify single failure modes and to quantify these modes; the analysis needs be no more elaborate than is necessary for these objectives.

4. Failure Mode Effect and Criticality Analysis (FMECA)

Failure Mode Effect and Criticality Analysis (FMECA), is essentially similar to a Failure Mode and Effects Analysis in which the criticality of the failure is analyzed in greater detail, and assurances and controls are described for limiting the likelihood of such failures. Although FMECA is not an optimal method for detecting hazards, it is frequently used in the course of a system safety analysis. The four fundamental facets of such an approach are (1) Fault Identification, (2) Potential Effects of the Fault, (3) Existing or Projected Compensation and/or Control, and (4) Summary of Findings. These four facets generally appear as column headings in an FMECA layout. Column 1 identifies the possible hazardous condition. Column 2 explains why this condition is a problem. Column 3 describes what has been done to compensate for or to control the condition. Finally, Column 4 states whether the situation is under control or whether further steps should be taken.

At this point the reader should be warned of a most hazardous pitfall that is present to a greater or lesser extent in all these inductive techniques: the potential of mistaking form for substance. If the project becomes simply a matter of filling out forms instead of conducting a proper analysis, the exercise will be completely futile. For this reason it might be better for the analyst not to restrict himself to any prepared formalism. Another point: if the system is at all complex, it is foolhardy for a single analyst to imagine that he alone can conduct a correct and comprehensive survey of all system faults and their effects on the system. These techniques call for a well-coordinated team approach.

5. Preliminary Hazard Analysis (PHA)

The techniques described so far have been, for the most part, system oriented, i.e., the effects are faults on the system operation. The subject of this section Preliminary Hazard Analysis (PHA), is a method for assessing the potential hazards posed, to plant personnel and other humans, by the system.

The objectives of a PHA are to identify potential hazardous conditions inherent within the system and to determine the significance or criticality of potential accidents that might arise. A PHA study should be conducted as early in the product development stage as possible. This will permit the early development of design and procedural safety requirements for controlling these hazardous conditions, thus eliminating costly design changes later on.

The first step in a PHA is to identify potentially hazardous elements or components within the system. This process is facilitated by engineering experience, the exercise of engineering judgment, and the use of numerous checklists that have been developed from time to time. The second step in a PHA is the identification of those events that could possibly transform specific hazardous conditions into potential accidents. Then the seriousness of these potential accidents is assessed to determine whether preventive measures should be taken.

Various columnar formats have been developed to facilitate the PHA process. Perhaps the simplest goes something like this:

Column (1)—Component/subsystem and hazard modes
Column (2)—Possible effects
Column (3)—Compensation and control
Column (4)—Findings and remarks

6. Fault Hazard Analysis (FHA)

Another method, Fault Hazard Analysis (FHA), was developed as a special purpose tool for use on projects involving many organizations, one of whom is supposed to act as integrator. This technique is especially valuable for detecting faults that cross organizational interfaces. It was first used to good purpose in the Minuteman III program.

A typical FHA form uses several columns as follows:

Column (1)—Component identification
Column (2)—Failure probability
Column (3)—Failure modes (identify all possible modes)
Column (4)—Percent failures by mode
Column (5)—Effect of failure (traced up to some relevant interface)
Column (6)—Identification of upstream component that could command or initiate the fault in question
Column (7)—Factors that could cause secondary failures (including threshold levels). This column should contain a listing of those operational or environmental variables to which the component is sensitive.
Column (8)—Remarks

The FHA is generally like an FMEA or FMECA with the addition of the extra information given in Columns 6 and 7.

As will become apparent in later chapters, Columns 6 and 7 have special significance for the fault tree analyst.

7. Double Failure Matrix (DFM)

The previous techniques concerned themselves with the effects of single failures. An inductive technique that also considers the effects of double failures is the Double Failure Matrix (DFM); its use is feasible only for relatively noncomplex systems. In order to illustrate its use, we must first discuss various ways in which faults may be categorized. A basic categorization which is related to that given in MIL STD 882 and modified for these discussions is as shown in Table II-2.

Table II-2. Fault Categories and Corresponding System Effects

Fault Category	Effect on System
I	Negligible
II	Marginal
III	Critical
IV	Catastrophic

It is desirable to give more complete definitions of the system effects:

(I) Negligible—loss of function that has no effect on system.

(II) Marginal—this fault will degrade the system to some extent but will not cause the system to be unavailable; for example, the loss of one of two redundant pumps, either of which can perform a required function.

(III) Critical—this fault will completely degrade system performance; for example, the loss of a component which renders a safety system unavailable.

(IV) Catastrophic—this fault will produce severe consequences which can involve injuries or fatalities; for example, catastrophic pressure vessel failure.

The categorization will depend on the conditions assumed to exist previously, and the categorizations can change as the assumed conditions change. For example, if one pump is assumed failed, then the failure of a second redundant pump is a critical failure.

The above crude categorizations can be refined in many ways. For example, on the NERVA project, six fault categories were defined as shown in Table II-3.

Table II-3. Fault Categories for NERVA Project

Fault Category	Effect on System
I	Negligible
IIA	A second fault event causes a transition into Category III (Critical)
IIB	A second fault event causes a transition into Category IV (Catastrophic)
IIC	A system safety problem whose effect depends upon the situation (e.g., the failure of all backup onsite power sources, which is no problem as long as primary, offsite power service remains on)
III	A critical failure and mission must be aborted
IV	A catastrophic failure

To illustrate the concept of DFM, consider the simple subsystem shown in Figure II-2. In this figure, the block valves can operate only as either fully open or fully closed, whereas the control valves are proportional valves which may be partially open or partially closed.

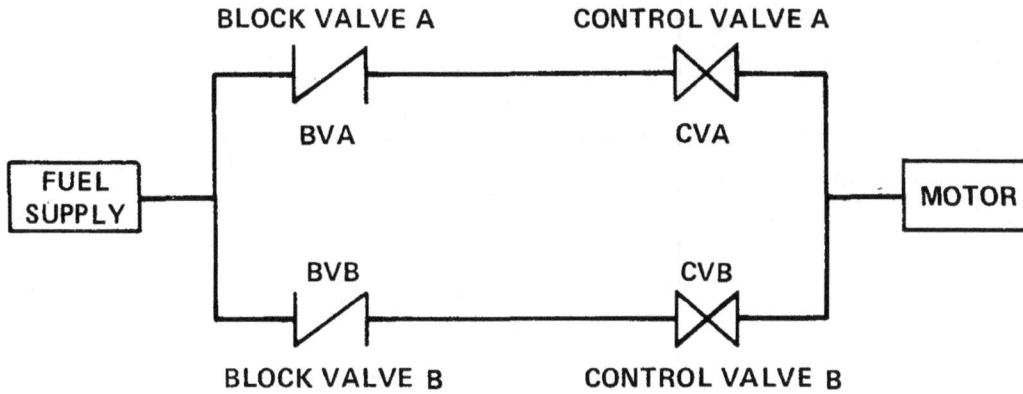

Figure II-2. Fuel System Schematic

Let us define two fault states for this system and categorize them as follows:

Fault State	Category
No flow when needed	IV
Flow cannot be shut off	III

We now proceed to consider all possible component failures and their fault categories. For instance, if Block Valve A (BVA) is failed open we have Category IIA because, if Control Valve A (CVA) is also failed open, we cascade into Category III. If BVA is failed closed we have Category IIB because, if either BVB or CVB is also failed closed, we cascade into Category IV. This type of analysis is conveniently systematized in the Double Failure Matrix shown in Table II-4.

For illustrative purposes we have filled in the entire matrix; for a first-order analysis we would be concerned only with the main diagonal terms, to wit, the single failure states. Note that if BVA is failed open, there is only one way in which a second failure can cascade us into Category III; namely, CVA must be failed open too. In contrast, if BVA is failed closed, we can cascade into Category IV if either BVB or CVB is also failed closed which is why "Two Ways" is given in Table II-4. Similar considerations apply to the single failures of CVA, BVB and CVB and this important additional information has been displayed in the principal diagonal cells of the matrix.

Now concentrating only on single failures, we can conduct a hazard category count as the following table shows:

Hazard Category	Number of Ways of Occurring
IIA	4
IIB	8

Table II-4. Fuel System Double Failure Matrix

		BVA		CVA		BVB		CVB	
		Open	Closed	Open	Closed	Open	Closed	Open	Closed
BVA	Open	(One Way) IIA	╳	III	IIB	IIA	IIA or IIB	IIA	IIA
	Closed	╳	(Two Ways) IIB	IIB	IIB	IIA or IIB	IV	IIA or IIB	IV
CVA	Open	III	IIB	(One Way) IIA	╳	IIA	IIA or IIB	IIA	IIA or IIB
	Closed	IIB	IIB	╳	(Two Ways) IIB	IIA	IV	IIA or IIB	IV
BVB	Open	IIA	IIA or IIB	IIA	IIA or IIB	(One Way) IIA	╳	III	IIB
	Closed	IIA or IIB	IV	IIA or IIB	IV	╳	(Two Ways) IIB	IIB	IIB
CVB	Open	IIA	IIA or IIB	IIA	IIA or IIB	III	IIB	(One Way) IIA	╳
	Closed	IIA or IIB	IV	IIA or IIB	IV	IIB	IIB	╳	(Two Ways) IIB

How can this information be used? One application would be a description and subsequent review of how these hazard categories are controlled or are insured against. Another application would be a comparison between the configuration of valves shown in Figure II-2 and an alternative design, for instance the configuration shown in Figure II-3.

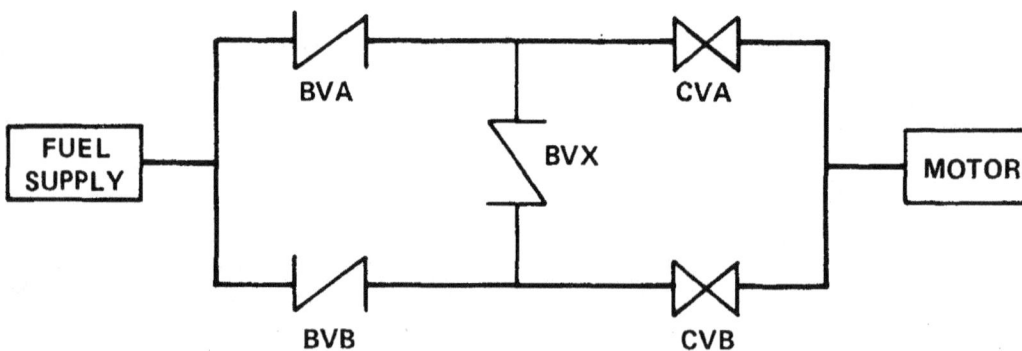

Figure II-3. Alternative Fuel System Schematic

For brevity let us refer to the system of Figure II-2 as "Configuration I" and that of Figure II-3 as "Configuration II." For Configuration II we naturally define the same system fault states as for Configuration I; namely, "no flow when needed" is

Category IV and "flow cannot be shut off" is Category III. We can now pose the following question: "Which configuration is the more desirable with respect to the relative number of occurrences of the various hazard categories that we have defined?" The appropriate Double Failure Matrix for Configuration II is shown in Table II-5.

Table II-5. Alternative Fuel System Double Failure Matrix

		BVA		CVA		BVB		CVB		BVX	
		Open	Closed	Open	Closed	Open	Closed	Open	Closed	O	C
BVA	Open	IIA (One Way)									
	Closed		IIB (One Way)								
CVA	Open			IIA (One Way)							
	Closed				IIB (One Way)						
BVB	Open					IIA (One Way)					
	Closed						IIB (One Way)				
CVB	Open							IIA (One Way)			
	Closed								IIB (One Way)		
BVX	Open									I	
	Closed										I

In this case we have filled in only the principal diagonal cells which correspond to single failure states. We see that if BVX is failed closed, Configuration II becomes essentially identical to Configuration I, and if BVX is failed open, we have a pipe connecting the two main flow channels.

Now, concentrating only on the single failure states, we can conduct another hazard category count for Configuration II. The results are shown in the following table:

Hazard Category	Number of Ways of Occurring
IIA	4
IIB	4

Comparing the two configurations, we see that they are the same from the standpoint of cascading into Category III but that Configuration II has approximately one-half as many ways to cascade into Category IV. Therefore, using this criterion, Configuration II is the better design. Where differences are not as obvious, more formal analysis approaches may also be used for additional information (these approaches will be discussed in the later sections).

8.   Success Path Models

We have been and will be discussing failures. Instead of working in "failure space" we can equivalently work in "success space." We give a brief example of the equivalence and then return to our failure space approach.

Consider the configuration of two valves in parallel shown in Figure II-4. This system may be analyzed either by a consideration of single failures (the probabilities of multiple failures are deemed negligible) or by a consideration of "success paths." Let us take up the former first.

System requirements are as follows:

(1) The operation involves two phases;

(2) At least one valve must open for each phase;

(3) Both valves must be closed at the end of each phase.

The two relevant component failure modes are: valve fails to open on demand, and valve fails to close on demand. For purposes of the analysis, let us assume the following numbers:

P (valve does not open) = 1×10^{-4} for each phase

P (valve does not close) = 2×10^{-4} for each phase

where the symbol "P" denotes probability. The valves are assumed to be identical.

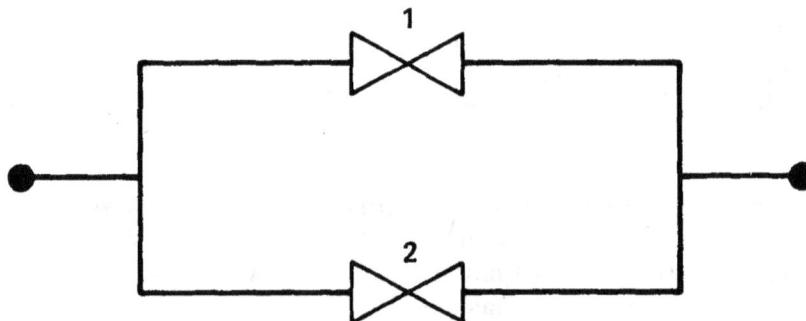

Figure II-4. Redundant Configuration of 2 Valves

The single failure analysis of the system can be tabulated as in Table II-6.

Table II-6. Single Failure Analysis for Redundant Valve Configuration

COMPONENT	FAILURE MODE	FAILURE EFFECT	PROBABILITY OF OCCURRENCE (F)
Valve # 1	Failure to open Failure to close	— System failure	4×10^{-4} (either phase)
Valve # 2	Failure to open Failure to close	— System failure	4×10^{-4} (either phase)

The system failure probability = 8×10^{-4}.

Now let us see whether we can duplicate this result by considering the possible successes. There are three identifiable success paths which we can specify both verbally and schematically. If R_O^i denotes "valve i opens successfully," and R_C^i denotes "valve i closes successfully," and P (Path i) denotes the success probability associated with the ith success path, we have the following:

Path 1: Both valves function properly for both cycles.

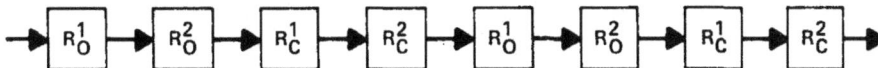

$$P(\text{Path 1}) = (R_O R_C)^4$$

Path 2: One valve fails to open on the first cycle but the other valve functions properly for both cycles.

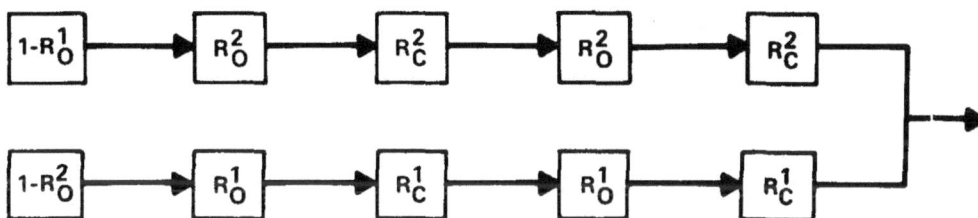

$$P(\text{Path 2}) = 2(1 - R_O)(R_O R_C)^2$$

Path 3: One valve fails to open on the second cycle but the other valve functions properly for both cycles.

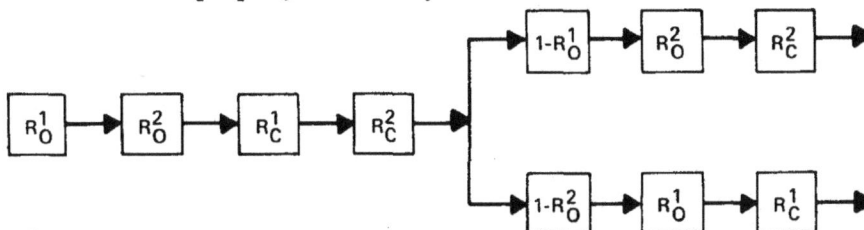

$$P(\text{Path 3}) = 2(1 - R_O)(R_O R_C)^3$$

Numerically, system reliability is given by

$$
\begin{aligned}
R_{SYSTEM} &= (R_O R_C)^4 + 2(1 - R_O)(R_O R_C)^2 + 2(1 - R_O)(R_O R_C)^3 \\
&= 0.99880027 + 0.00019988 + 0.00019982 \\
&= 0.99919997 \cong 1 - 8 \times 10^{-4}
\end{aligned}
$$

which is essentially the same result as before but it can be seen that the failure approach is considerably less laborious.

9. Conclusions

Although the various inductive methods that we have discussed can be elaborated to almost any desirable extent, in actual practice they generally play the role of "overview" methods and, on many occasions, this is all that is necessary. For any reasonably complex system, the identification of all component failure modes will be a laborious, and probably unnecessary, process. Worse yet, the identification of all possible combinations of component failure modes will be a truly Herculean task. In general, it is a waste of time to bother with failure effects (single or in combination) that have little or no effect on system operation or whose probabilities of occurrence are entirely negligible. Thus, in all of these analyses the consequences of a certain event must be balanced against its likelihood of occurrence.

CHAPTER III – FAULT TREE ANALYSIS –
BASIC CONCEPTS

1. Orientation

In Chapter I we introduced two approaches to system analysis: inductive and deductive. Chapter II described the major inductive methods. Chapter III presents the basic concepts and definitions necessary for an understanding of the deductive Fault Tree Analysis approach, which is the subject of the remainder of this text.

2. Failure vs. Success Models

The operation of a system can be considered from two standpoints: we can enumerate various ways for system success, or we can enumerate various ways for system failure. We have already seen an example of this in Chapter II, section 8. Figure III-1 depicts the Failure/Success space concept.

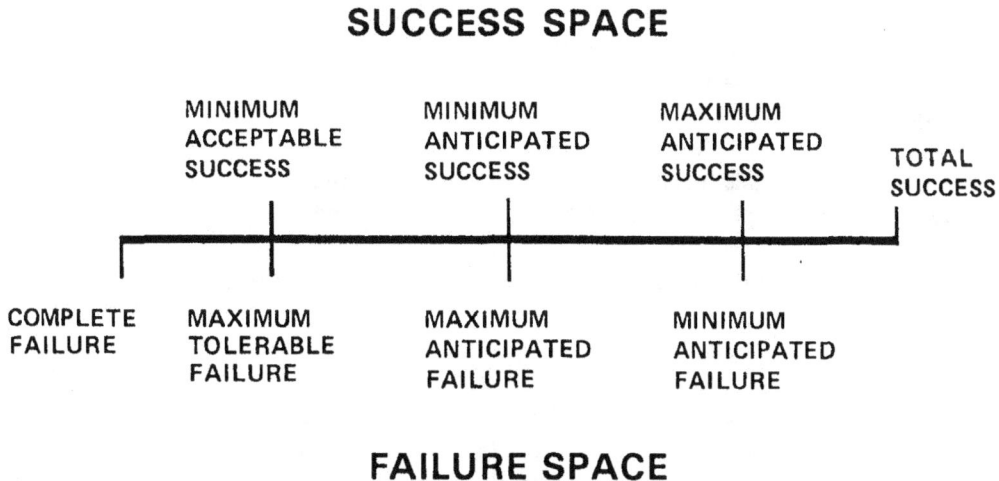

SUCCESS SPACE

| MINIMUM ACCEPTABLE SUCCESS | MINIMUM ANTICIPATED SUCCESS | MAXIMUM ANTICIPATED SUCCESS | TOTAL SUCCESS |

| COMPLETE FAILURE | MAXIMUM TOLERABLE FAILURE | MAXIMUM ANTICIPATED FAILURE | MINIMUM ANTICIPATED FAILURE |

FAILURE SPACE

Figure III-1. The Failure Space–Success Space Concept

It is interesting to note that certain identifiable points in success space coincide with certain analogous points in failure space. Thus, for instance, "maximum anticipated success" in success space can be thought of as coinciding with "minimum anticipated failure" in failure space. Although our first inclination might be to select the optimistic view of our system—success—rather than the pessimistic one—failure—, we shall see that this is not necessarily the most advantageous one.

From an analytical standpoint, there are several overriding advantages that accrue to the failure space standpoint. First of all, it is generally easier to attain concurrence on what constitutes failure than it is to agree on what constitutes success. We may desire an airplane that flies high, travels far without refueling, moves fast and carries a big load. When the final version of this aircraft rolls off the production line, some of these features may have been compromised in the course of making the usual trade-offs. Whether the vehicle is a "success" or not may very well be a matter of controversy. On the other hand, if the airplane crashes in flames, there will be little argument that this event constitutes system failure.

"Success" tends to be associated with the efficiency of a system, the amount of output, the degree of usefulness, and production and marketing features. These characteristics are describable by continuous variables which are not easily modeled in terms of simple discrete events, such as "valve does not open" which characterizes the failure space (partial failures, i.e., a valve opens partially, are also difficult events to model because of their continuous possibilities). Thus, the event "failure," in particular, "complete failure," is generally easy to define, whereas the event, "success," may be much more difficult to tie down. This fact makes the use of failure space in analysis much more valuable than the use of success space.

Another point in favor of the use of failure space is that, although theoretically the number of ways in which a system can fail and the number of ways in which a system can succeed are both infinite, from a practical standpoint there are generally more ways to success than there are to failure. Thus, purely from a practical point of view, the size of the population in failure space is less than the size of the population in success space. In analysis, therefore, it is generally more efficient to make calculations on the basis of failure space.

We have been discussing why it is more advantageous for the analyst to work in failure space as opposed to success space. Actually all that is necessary is to demonstrate that consideration of failure space allows the analyst to get his job done, and this, indeed, has been shown many times in the past. The drawing of tree diagrams for a complex system is an expensive and time-consuming operation. When failures are considered, it may be necessary to construct only one or two system models such as fault trees, which cover all the significant failure modes. When successes are considered, it may become necessary to construct several hundred system models covering various definitions of success. A good example of the parsimony of events characteristic of failure space is the Minuteman missile analysis. Only three fault trees were drawn corresponding to the three undesired events: inadvertent programmed launch, accidental motor ignition, and fault launch. It was found that careful analysis of just these three events involved a complete overview of the whole complex system.

To help fix our ideas, it may be helpful to subject some everyday occurrence (a man driving to his office) to analysis in failure space (see Figure III-2).

The "mission" to which Figure III-2 refers is the transport of Mr. X by automobile from his home to his office. The desired arrival time is 8:30, but the mission will be considered marginally successful if Mr. X arrives at his office by 9:00. Below "minimum anticipated failure" lie a number of possible incidents that constitute minor annoyances, but which do not prevent Mr. X from arriving at the desired time. Arrival at 9:00 is labeled "maximum anticipated failure." Between this point and "minimum anticipated failure" lie a number of occurrences that cause Mr. X's arrival time to be delayed half an hour or less. It is perhaps reasonable to let the point "maximum tolerable failure" coincide with some accident that causes some damage to the car and considerable delay but no personal injury. Above this point lie incidents of increasing seriousness terminating in the ultimate catastrophe of death.

```
COMPLETE FAILURE ——————▶◯◀——— ACCIDENT
                                  (DEATH OR CRIPPLING INJURY)

MAXIMUM TOLERABLE FAILURE——▶ │◀——— ACCIDENT
                              │       (CAR DAMAGED; NO PERSONAL INJURY)
                              │
                              │◀——— MINOR ACCIDENT
                              │
                              │◀——— FLAT TIRE
                              │
                              │◀——— WINDSHIELD WIPERS INOPERATIVE
                              │       (HEAVY RAIN)
                              │
                              │◀——— TRAFFIC JAM
                              │
MAXIMUM ANTICIPATED FAILURE——▶│◀——— ARRIVES AT 9:00
                              │
                              │◀——— WINDSHIELD WIPERS INOPERATIVE
                              │       (LIGHT RAIN)
                              │
                              │◀——— TRAFFIC CONGESTION
                              │
MINIMUM ANTICIPATED FAILURE——▶│◀——— ARRIVES AT 8:45
                              │
                              │◀——— LOST HUBCAP
                              │
                              │◀——— WINDSHIELD WIPERS INOPERATIVE
                              │       (CLEAR WEATHER)
                              │
TOTAL SUCCESS ——————▶◯◀——— ARRIVES AT 8:30
                                  (NO DIFFICULTIES WHATSOEVER)
```

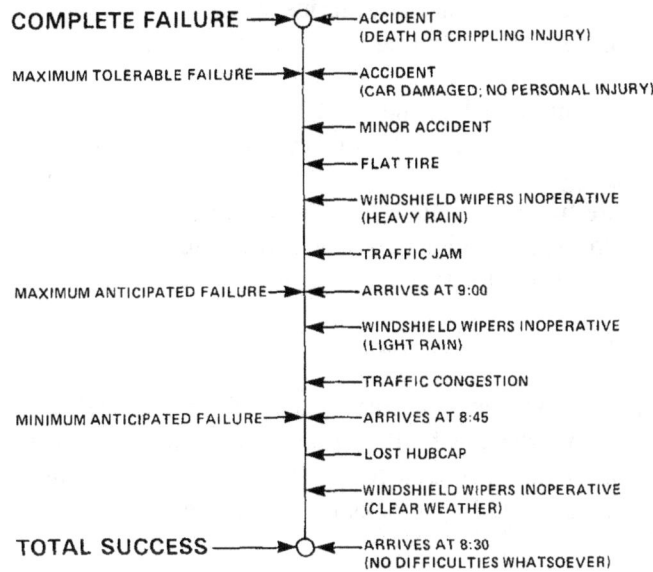

Figure III-2. Use of Failure Space in Transport Example

Note that an event such as "windshield wipers inoperative" will be positioned along the line according to the nature of the environment at that time.

A chart such as Figure III-2 might also be used to pinpoint events in, for example, the production of a commercial airliner. The point "minimum anticipated failure" would correspond to the attainment of all specifications and points below that would indicate that some of the specifications have been _more_ than met. The point "maximum anticipated failure" would correspond to some trade-off point at which all specifications had not been met but the discrepancies were not serious enough to degrade the saleability of the airplane in a material way. The point "maximum tolerable failure" corresponds to the survival point of the company building the aircraft. Above that point, only intolerable catastrophes occur. Generally speaking, Fault Tree Analysis addresses itself to the identification and assessment of just such catastrophic occurrences and complete failures.

3.   The Undesired Event Concept

Fault tree analysis is a deductive failure analysis which focuses on one particular undesired event and which provides a method for determining causes of this event. The undesired event constitutes the top event in a fault tree diagram constructed for the system, and generally consists of a complete, or catastrophic failure as mentioned above. Careful choice of the top event is important to the success of the analysis. If it is too general, the analysis become unmanageable; if it is too specific, the analysis does not provide a sufficiently broad view of the system. Fault tree analysis can be an expensive and time-consuming exercise and its cost must be measured against the cost associated with the occurrence of the relevant undesired event.

We now give some examples of top events that might be suitable for beginning a fault tree analysis:

(a) Catastrophic failure of a submarine while the submarine is submerged. In the analysis we might separate "failure under hostile attack" from "failure under routine operation."

(b) Crash of commercial airliner with loss of several hundred lives.

(c) No spray when demanded from the containment spray injection system in a nuclear reactor.

(d) Premature full-scale yield of a nuclear warhead.

(e) Loss of spacecraft and astronauts in the space exploration program.

(f) Automobile does not start when ignition key is turned.

4. Summary

In this chapter we have discussed the "failure space" and "undesired event" concepts which underlie the fault tree approach. In the next chapter we will define Fault Tree Analysis and proceed to a careful definition of the gates and fault events which constitute the building blocks of a fault tree.

CHAPTER IV – THE BASIC ELEMENTS OF A FAULT TREE

1. The Fault Tree Model

A fault tree analysis can be simply described as an analytical technique, whereby an undesired state of the system is specified (usually a state that is critical from a safety standpoint), and the system is then analyzed in the context of its environment and operation to find all credible ways in which the undesired event can occur. The fault tree itself is a graphic model of the various parallel and sequential combinations of faults that will result in the occurrence of the predefined undesired event. The faults can be events that are associated with component hardware failures, human errors, or any other pertinent events which can lead to the undesired event. A fault tree thus depicts the logical interrelationships of basic events that lead to the undesired event—which is the top event of the fault tree.

It is important to understand that a fault tree is not a model of all possible system failures or all possible causes for system failure. A fault tree is tailored to its top event which corresponds to some particular system failure mode, and the fault tree thus includes only those faults that contribute to this top event. Moreover, these faults are not exhaustive—they cover only the most credible faults as assessed by the analyst.

It is also important to point out that a fault tree is not in itself a quantitative model. It is a qualitative model that can be evaluated quantitatively and often is. This qualitative aspect, of course, is true of virtually all varieties of system models. The fact that a fault tree is a particularly convenient model to quantify does not change the qualitative nature of the model itself.

A fault tree is a complex of entities known as "gates" which serve to permit or inhibit the passage of fault logic up the tree. The gates show the relationships of events needed for the occurrence of a "higher" event. The "higher" event is the "output" of the gate; the "lower" events are the "inputs" to the gate. The gate symbol denotes the type of relationship of the input events required for the output event. Thus, gates are somewhat analogous to switches in an electrical circuit or two valves in a piping layout. Figure IV-1 shows a typical fault tree.

2. Symbology—The Building Blocks of the Fault Tree

A typical fault tree is composed of a number of symbols which are described in detail in the remaining sections of this chapter and are summarized for the reader's convenience in Table IV-1.

PRIMARY EVENTS

The primary events of a fault tree are those events, which, for one reason or another, have not been further developed. These are the events for which

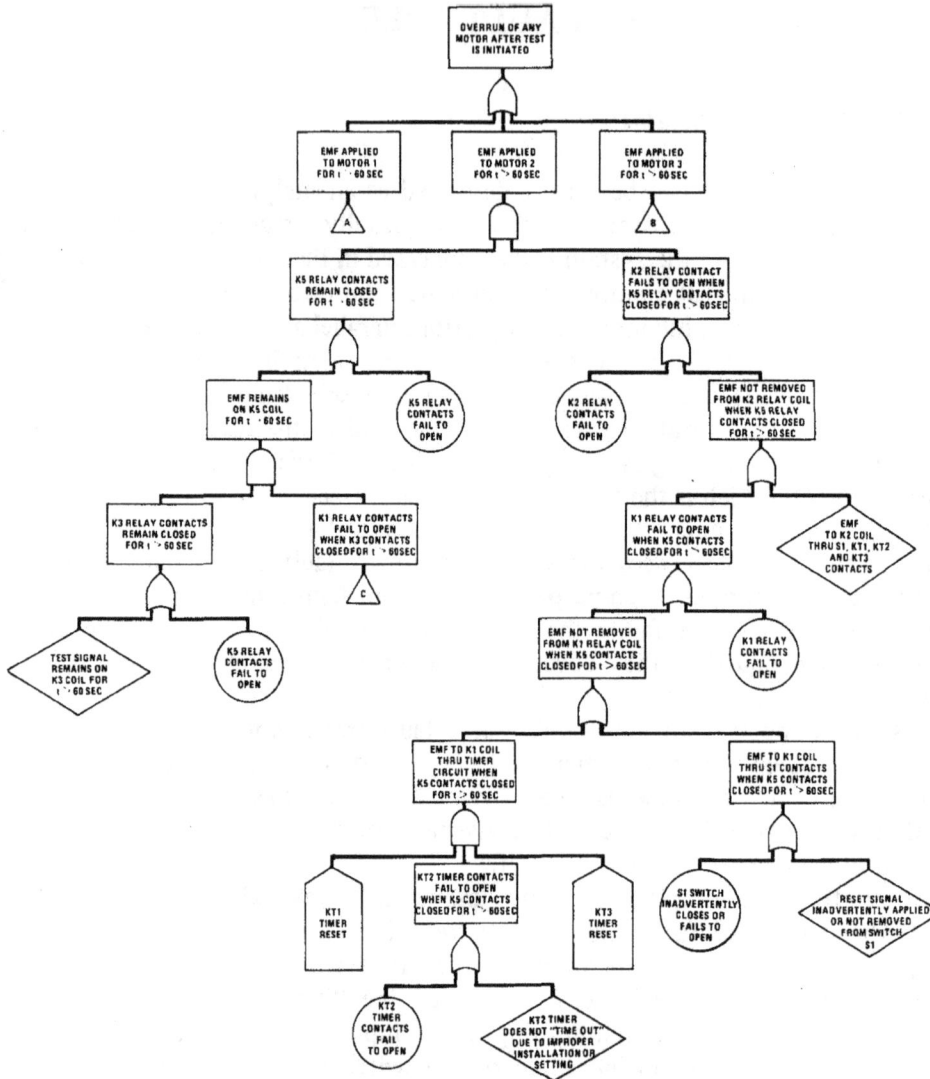

Figure IV-1. Typical Fault Tree

probabilities will have to be provided if the fault tree is to be used for computing the probability of the top event. There are four types of primary events. These are:

The Basic Event

The circle describes a basic initiating fault event that requires no further development. In other words, the circle signifies that the appropriate limit of resolution has been reached.

Table IV-1. Fault Tree Symbols

PRIMARY EVENT SYMBOLS

BASIC EVENT — A basic initiating fault requiring no further development

CONDITIONING EVENT — Specific conditions or restrictions that apply to any logic gate (used primarily with PRIORITY AND and INHIBIT gates)

UNDEVELOPED EVENT — An event which is not further developed either because it is of insufficient consequence or because information is unavailable

EXTERNAL EVENT — An event which is normally expected to occur

INTERMEDIATE EVENT SYMBOLS

INTERMEDIATE EVENT — A fault event that occurs because of one or more antecedent causes acting through logic gates

GATE SYMBOLS

AND — Output fault occurs if all of the input faults occur

OR — Output fault occurs if at least one of the input faults occurs

EXCLUSIVE OR — Output fault occurs if exactly one of the input faults occurs

PRIORITY AND — Output fault occurs if all of the input faults occur in a specific sequence (the sequence is represented by a CONDITIONING EVENT drawn to the right of the gate)

INHIBIT — Output fault occurs if the (single) input fault occurs in the presence of an enabling condition (the enabling condition is represented by a CONDITIONING EVENT drawn to the right of the gate)

TRANSFER SYMBOLS

TRANSFER IN — Indicates that the tree is developed further at the occurrence of the corresponding TRANSFER OUT (e.g., on another page)

TRANSFER OUT — Indicates that this portion of the tree must be attached at the corresponding TRANSFER IN

The Undeveloped Event

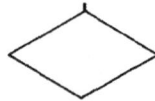

The diamond describes a specific fault event that is not further developed, either because the event is of insufficient consequence or because information relevant to the event is unavailable.

The Conditioning Event

The ellipse is used to record any conditions or restrictions that apply to any logic gate. It is used primarily with the INHIBIT and PRIORITY AND-gates.

The External Event

The house is used to signify an event that is normally expected to occur: e.g., a phase change in a dynamic system. Thus, the house symbol displays events that are not, of themselves, faults.

INTERMEDIATE EVENTS

An intermediate event is a fault event which occurs because of one or more antecedent causes acting through logic gates. All intermediate events are symbolized by rectangles.

GATES

There are two basic types of fault tree gates: the OR-gate and the AND-gate. All other gates are really special cases of these two basic types. With one exception, gates are symbolized by a shield with a flat or curved base.

The OR-Gate

The OR-gate is used to show that the output event occurs only if one or more of the input events occur. There may be any number of input events to an OR-gate.

Figure IV-2 shows a typical two-input OR-gate with input events A and B and output event Q. Event Q occurs if A occurs, B occurs, or both A and B occur.

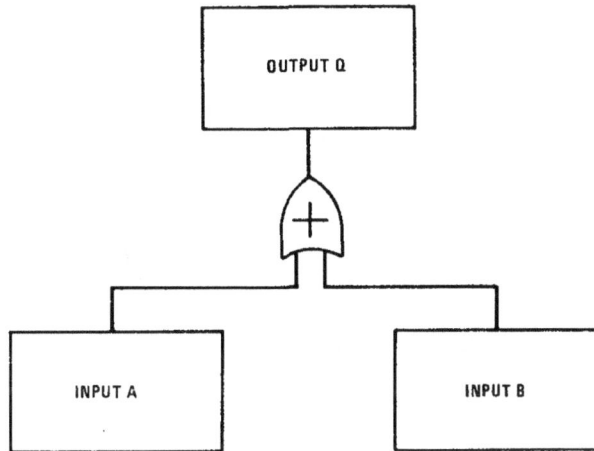

Figure IV-2. The OR-Gate

It is important to understand that <u>causality never passes through an OR-gate.</u> That is, for an OR-gate, the input faults are never the causes of the output fault. Inputs to an OR-gate are identical to the output but are more specifically defined as to cause. Figure IV-3 helps to clarify this point.

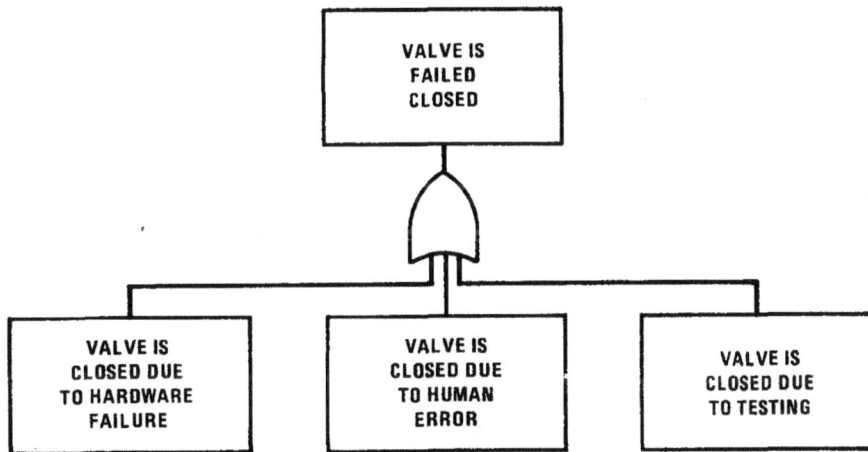

Figure IV-3. Specific Example of the OR-Gate

Note that the subevents in **Figure IV-3** can be further developed; for instance, see Figure IV-4.

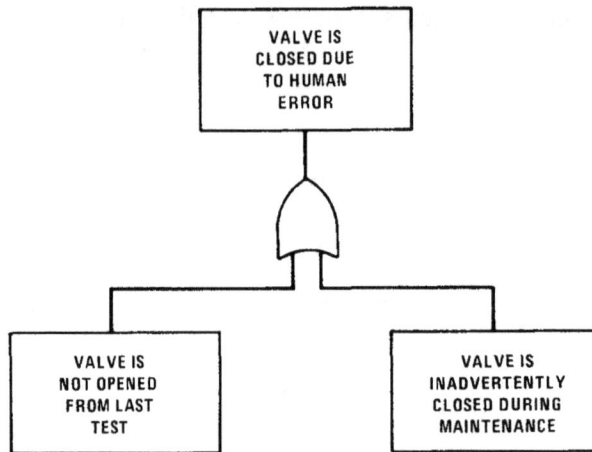

Figure IV-4. OR-Gate for Human Error

However, the event

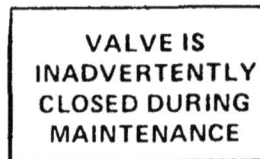

is still a restatement of the output event of the first OR-gate

with regard to a specific cause.

One way to detect improperly drawn fault trees is to look for cases in which causality passes through an OR-gate. This is an indication of a missing AND-gate (see following definition) and is a sign of the use of improper logic in the conduct of the analysis.

The AND-Gate

The AND-gate is used to show that the output fault occurs only if all the input faults occur. There may be any number of input faults to an AND-gate. Figure IV-5 shows a typical two-input AND-gate with input events A and B, and output event Q. Event Q occurs only if events A and B both occur.

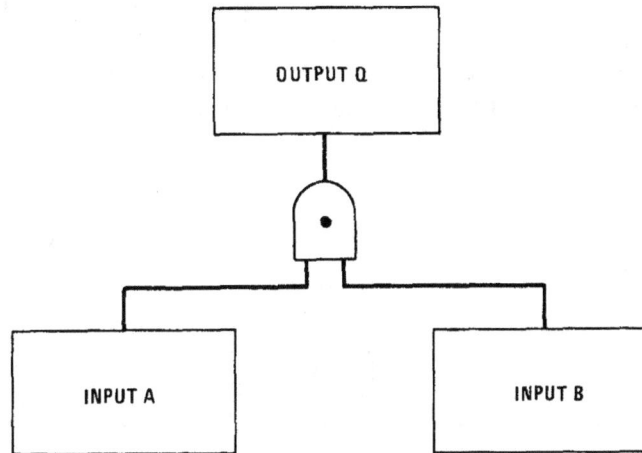

Figure IV-5. The AND-Gate

In contrast to the OR-gate the AND-gate does specify a causal relationship between the inputs and the output, i.e., the input faults collectively represent the cause of the output fault. The AND-gate implies nothing whatsoever about the antecedents of the input faults. An example of an AND-gate is shown in Figure IV-6. A failure of both diesel generators and of the battery will result in a failure of all onsite DC power.

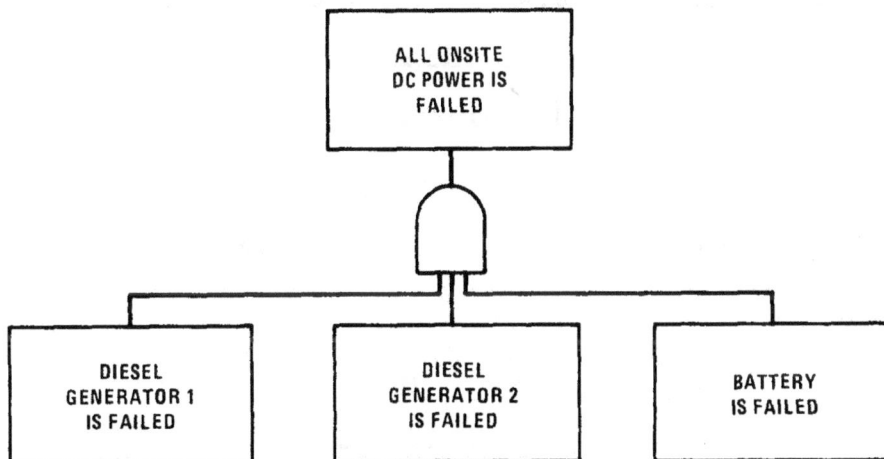

Figure IV-6. Specific Example of an AND-Gate

When describing the events input to an AND-gate, any dependencies must be incorporated in the event definitions if the dependencies affect the system logic. Dependencies generally exist when the failure "changes" the system. For example, when the first failure occurs (e.g., input A of Figure IV-5), the system may automatically switch in a standby unit. The second failure, input B of Figure IV-5, is now analyzed with the standby unit assumed to be in place. In this case, input B of Figure IV-5 would be more precisely defined as "input B given the occurrence of A."

The variant of the AND-gate shown in Figure IV-7 explicitly shows the dependencies and is useful for those situations when the occurrence of one of the faults alters the operating modes and/or stress levels in the system in a manner affecting the occurrence mechanism of the other fault.

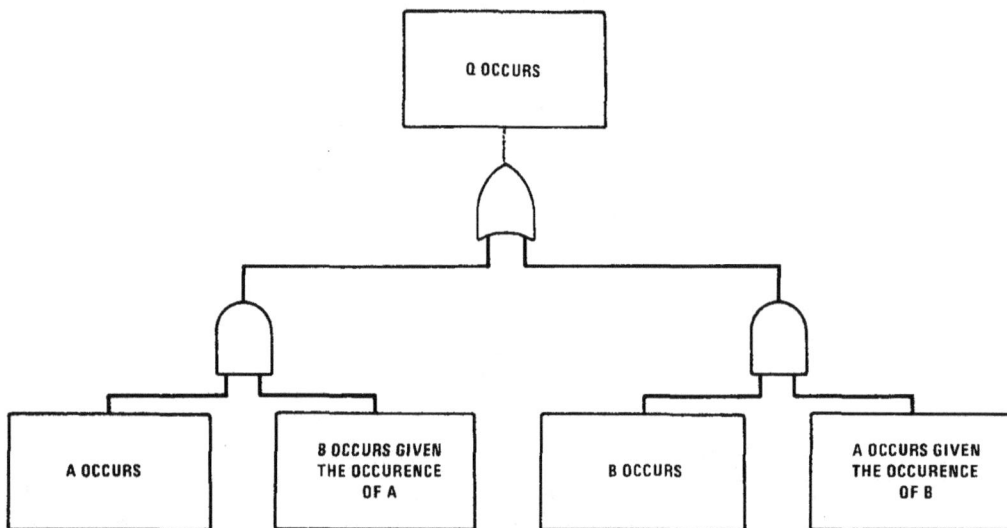

Figure IV-7. AND-Gate Relationship with Dependency Explicitly Shown

That is, the subtree describing the mechanisms or antecedent causes of the event

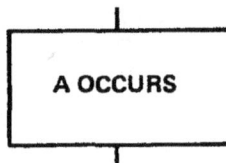

will be <u>different</u> from the subtree describing the mechanisms for the event.

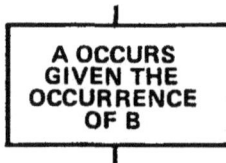

For multiple inputs to an AND-gate with dependencies affecting system logic among the input events, the "givens" must incorporate all preceding events.

The INHIBIT-Gate

The INHIBIT-gate, represented by the hexagon, is a special case of the AND-gate. The output is caused by a single input, but some qualifying condition must be satisfied before the input can produce the output. The condition that must exist is the conditional input. A description of this conditional input is spelled out within an ellipse drawn to the right of the gate. Figure IV-8 shows a typical INHIBIT-gate with input A, conditional input B and output Q. Event Q occurs only if input A occurs under the condition specified by input B.

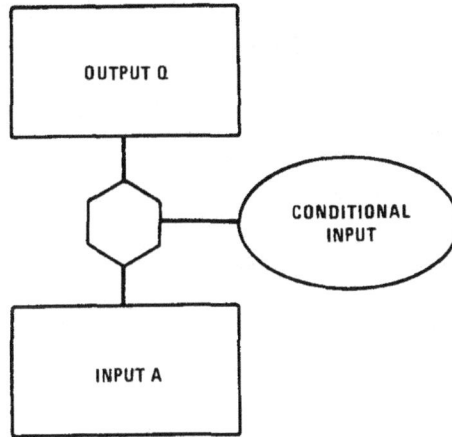

Figure IV-8. The INHIBIT-Gate

To clarify this concept, two examples are given below and are illustrated in Figure IV-9.

(a) Many chemical reactions go to completion only in the presence of a catalyst. The catalyst does not participate in the reaction, but its presence is necessary.

(b) If a frozen gasoline line constitutes an event of interest, such an event can occur only when the temperature T is less than $T_{critical}$, the temperature at which the gasoline freezes. In this case the output event would be "frozen gasoline line," the input event would be "existence of low temperature," and the conditional input would be "$T < T_{critical}$."

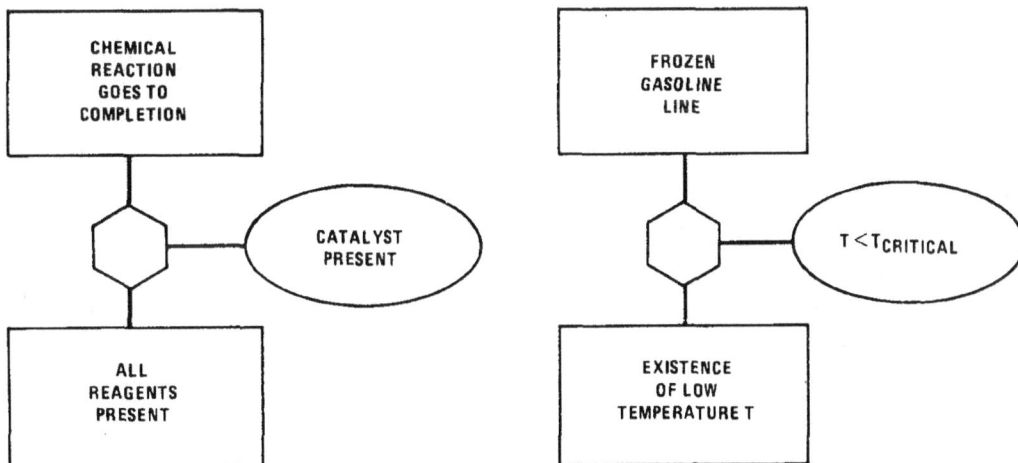

Figure IV-9. Examples of the INHIBIT-Gate

Occasionally, especially in the investigation of secondary failures (see Chapter V), another type of INHIBIT-gate depicted in Figure IV-10 is used.

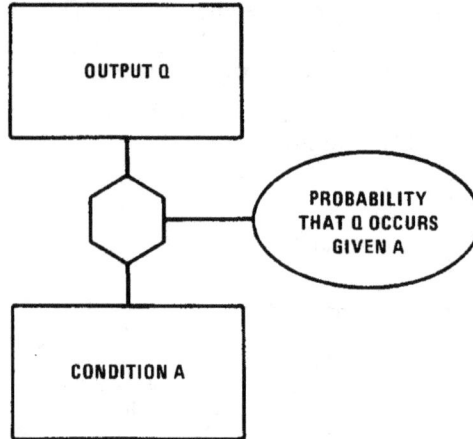

Figure IV-10. An Alternative Type of INHIBIT-Gate

In Figure IV-10, condition A is the necessary, but not always sufficient, single cause of output Q; i.e., for Q to occur we must have A, but just because A occurs it does not mean that Q follows inevitably. The portion of time Q occurs when A occurs is given in the conditional input ellipse.

The gates we have described above are the ones most commonly used and are now standard in the field of fault tree analysis. However, a few other special purpose gates are sometimes encountered.

The EXCLUSIVE OR-Gate

The EXCLUSIVE OR-gate is a special case of the OR-gate in which the output event occurs only if exactly one of the input events occur. Figure IV-11 shows two alternative ways of depicting a typical EXCLUSIVE OR-gate with two inputs.

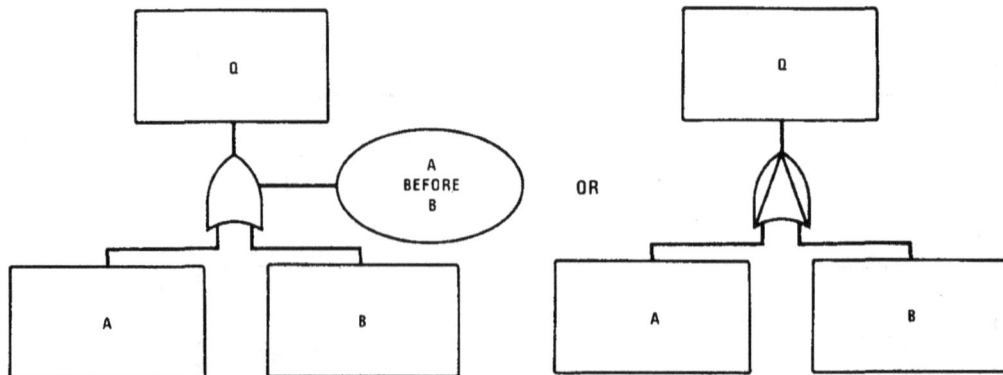

Figure IV-11. The EXCLUSIVE OR-Gate

The exclusive OR differs from the usual or inclusive OR in that the situation where both input events occur is precluded. Thus, the output event Q occurs if A occurs or B occurs, but not if both A and B occur. As we will see later in Chapter VII, the quantitative difference between the inclusive and exclusive OR-gates is generally so insignificant that the distinction is not usually necessary. In those special instances where the distinction is significant, it can be accounted for in the quantification phase.

The PRIORITY AND-Gate

The PRIORITY AND-gate is a special case of the AND-gate in which the output event occurs only if all input events occur in a specified ordered sequence. The sequence is usually shown inside an ellipse drawn to the right of the gate. In practice, the necessity of having a specific sequence is not usually encountered. Figure IV-12 shows two alternative ways of depicting a typical PRIORITY AND-gate.

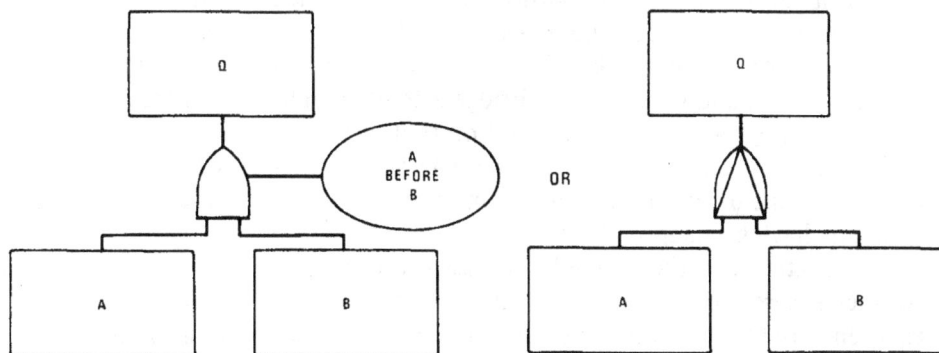

Figure IV-12. The PRIORITY AND-Gate

In Figure IV-11, the output event Q occurs only if both input events A and B occur with A occurring before B.

TRANSFER SYMBOLS

The triangles are introduced as transfer symbols and are used as a matter of convenience to avoid extensive duplication in a fault tree. A line from the apex of the triangle denotes a "transfer in," and a line from the side, a "transfer out." A "transfer in" attached to a gate will link to its corresponding "transfer out." This "transfer out," perhaps on another sheet of paper, will contain a further portion of the tree describing input to the gate.

CHAPTER V – FAULT TREE CONSTRUCTION FUNDAMENTALS

In Chapter IV we defined and discussed the symbols which are the building blocks of a fault tree. In this chapter we cover the concepts which are needed for the proper selection and definition of the fault tree events, and thus, for the construction of the fault tree.

1. Faults vs. Failures

We first make a distinction between the rather specific word "failure" and the more general word "fault." Consider a relay. If the relay closes properly when a voltage is impressed across its terminals, we call this a relay "success." If, however, the relay fails to close under these circumstances, we call this a relay "failure." Another possibility is that the relay closes at the wrong time due to the improper functioning of some upstream component. This is clearly not a relay failure; however, untimely relay operation may well cause the entire circuit to enter an unsatisfactory state. We shall call an occurrence like this a "fault" so that, generally speaking, all failures are faults but not all faults are failures. Failures are basic abnormal occurrences, whereas faults are "higher order" events.

Consider next a bridge that is supposed to open occasionally to allow the passage of marine traffic. Suddenly, without warning, one leaf of the bridge flips up a few feet. This is not a bridge failure because it is supposed to open on command and it does. However, the event is a fault because the bridge mechanism responded to an untimely command issued by the bridge attendant. Thus, the attendant is part of this "system," and it was his untimely action that caused the fault.

In one of the earliest battles of the American Civil War, General Beauregard sent a message to one of his officers via mounted messenger #1. Some time later, the overall situation having changed, he sent out an amended message via mounted messenger #2. Still later, a further amended message was sent out via mounted messenger #3. All messengers arrived but not in the proper sequence. No failure occurred, but such an event could well have a deleterious effect on the progress of the battle—and, in this case, it did. Again, we have a fault event but not a failure event.

The proper definition of a fault requires a specification of not only what the undesirable component state is but also when it occurs. These "what" and "when" specifications should be part of the event descriptions which are entered into the fault tree.

2. Fault Occurrence vs. Fault Existence

In our discussion of the several varieties of fault tree gates, we have spoken of the occurrence of one or more of a set of faults or the occurrence of all of a set of faults. A fault may be repairable or not, depending on the nature of the system. Under conditions of no repair, a fault that occurs will continue to exist. In a repairable system a distinction must be made between the occurrence of a fault and its existence. Actually this distinction is of importance only in fault tree quantification (to be discussed in a later chapter). From the standpoint of constructing a fault tree we need concern ourselves only with the phenomenon of occurrence. This is tantamount to considering all systems as nonrepairable.

3.   Passive vs. Active Components

In most cases it is convenient to separate components into two types: passive and active (also called quasi-static and dynamic). A passive component contributes in a more or less static manner to the functioning of the system. Such a component may act as a transmitter of energy from place to place (e.g., a wire or bus-bar carrying current or a steam line transmitting heat energy), or it may act as a transmitter of loads (e.g., a structural member). To assess the operation of a passive component, we perform such tests as stress analysis, heat transfer studies, etc. Further examples of passive components are: pipes, bearings, journals, welds, and so forth.

An active component contributes in a more dynamic manner to the functioning of its parent system by modifying system behavior in some way. A valve which opens and closes, for example, modifies the system's fluid flow, and a switch has a similar effect on the current in an electrical circuit. To assess the operation of an active component, we perform parametric studies of operating characteristics and studies of functional interrelationships. Examples of active components are: relays, resistors, pumps, and so forth.

A passive component can be considered as the transmitter of a "signal." The physical nature of this "signal" may exhibit considerable variety; for example, it may be a current or force. A passive component may also be thought of as the "mechanism" (e.g., a wire) whereby the output of one active component becomes the input to a second active component. The failure of a passive component will result in the non-transmission (or, perhaps, partial transmission) of its "signal."

In contrast, an active component originates or modifies a signal. Generally, such a component requires an input signal or trigger for its output signal. In such cases the active component acts as a "transfer function," a term widely used in electrical and mathematical studies. If an active component fails, there may be no output signal or there may be an incorrect output signal.

As an example, consider a postman (passive component) who transmits a signal (letter) from one active component (sender) to another (receiver). The receiver will then respond in some way (provide an output) as a result of the message (input) that he has received.

From a numerical reliability standpoint, the important difference between failures of active components and failures of passive components is the difference in failure rate values. As shown in WASH-1400 [39], active component failures in general have failure rates above 1×10^{-4} per demand (or above 3×10^{-7} per hour) and passive component failures in general have failure rates below these values. In fact, the difference in reliability between the two types of components is, quite commonly, two to three orders of magnitude.

In the above, the definitions of active components and passive components apply to the main function performed by the component; and failures of the active component (or failures of the passive component) apply to the failure of that main function. (We could have, for example, "passive" failure modes of active components, e.g., valve rupture, if we attempted to classify specific failure modes according to the "active" or "passive" definition.)

4.   Component Fault Categories: Primary, Secondary, and Command

It is also useful for the fault tree analyst to classify faults into three categories: primary, secondary and command. A primary fault is any fault of a component that occurs in an environment for which the component is qualified; e.g., a pressure tank, designed to withstand pressures up to and including a pressure p_0, ruptures at some pressure $p \leqslant p_0$ because of a defective weld.

A secondary fault is any fault of a component that occurs in an environment for which it has not been qualified. In other words, the component fails in a situation which exceeds the conditions for which it was designed; e.g., a pressure tank, designed to withstand pressure up to and including a pressure p_0, ruptures under a pressure $p > p_0$.

Because primary and secondary faults are generally component failures, they are usually called primary and secondary failures. A command fault in contrast, involves the proper operation of a component but at the wrong time or in the wrong place; e.g., an arming device in a warhead train closes too soon because of a premature or otherwise erroneous signal origination from some upstream device.

5.   Failure Mechanism, Failure Mode, and Failure Effect

The definitions of system, subsystem, and component are relative, and depend upon the context of the analysis. We may say that a "system" is the overall structure being considered, which in turn consists of subordinate structures called "subsystems," which in turn are made up of basic building blocks called "components."

For example, in a pressurized water reactor (PWR), the Spray Injection System may consist of two redundant refueling spray subsystems which deliver water from the refueling water storage tank to the containment. Each of these subsystems in turn may consist of arrangements of valves, pumps, etc., which are the components. In a particular analysis, definitions of system, subsystem, and component are generally made for convenience in order to give hierarchy and boundaries to the problem.

In constructing a fault tree, the basic concepts of failure effects, failure modes, and failure mechanisms are important in determining the proper interrelationships among the events. When we speak of failure effects, we are concerned about why the particular failure is of interest, i.e., what are its effects (if any) on the system. When we detail the failure modes, we are specifying exactly what aspects of component failure are of concern. When we list failure mechanisms, we are considering how a particular failure mode can occur and also, perhaps, what are the corresponding likelihoods of occurrence. Thus, failure mechanisms produce failure modes which, in turn, have certain effects on system operation.

To illustrate these concepts consider a system that controls the flow of fuel to an engine. See Table V-1. The subsystem of interest consists of a valve and a valve actuator. We can classify various events which can occur as viewed from the system, subsystem, or component level. Some of the events are given in the left-hand column of the table below. For example, "valve unable to open" is a mechanism of subsystem failure, a mode of valve failure, and an effect of actuator failure.

Table V-1. Fuel Flow System Failure Analysis

Description of Event	System	Subsystem	Valve	Actuator
No flow from subsystem when required	Mechanism	Mode	Effect	
Valve unable to open		Mechanism	Mode	Effect
Binding of actuator stem			Mechanism	Mode
Corrosion of actuator stem				Mechanism

To make the mechanism-mode-effect distinction clearer, consider a simple system of a doorbell and its associated circuitry from the standpoints of the systems man, the subsystems man, and the component designer. The system is shown schematically in Figure V-1.

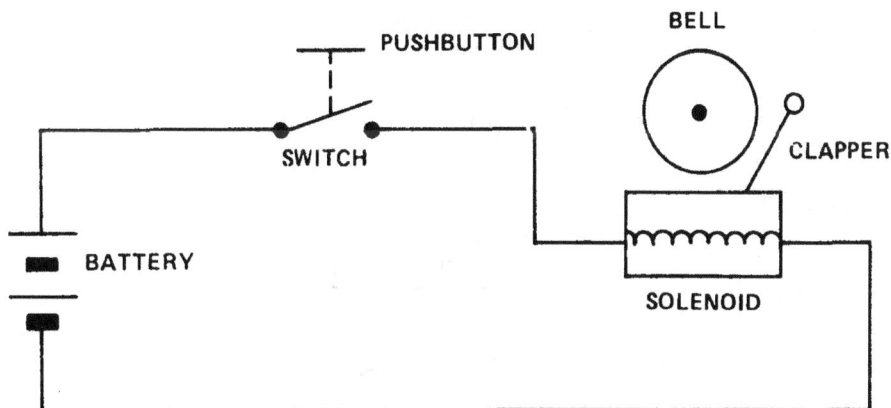

Figure V-1. Doorbell and Associated Circuitry

From the viewpoint of the systems man, the system failure modes are:
(1) Doorbell fails to ring when thumb pushes button.
(2) Doorbell inadvertently rings when button is not pushed.
(3) Doorbell fails to stop ringing when push button is released.
If the systems man now sat down and made a list of the failure mechanisms causing his failure modes, he would generate a list that corresponded to the failure modes of the subsystems man who actually procures the switch, bell-solenoid unit, battery, and wires. These are:
    (1) Switch — (a) fails to make contact (including an inadvertent open)
              (b) fails to break contact
              (c) inadvertently closes
    (2) Bell-solenoid unit—fails to ring when power is applied (includes failure to continue ringing with power applied)
    (3) Battery—low voltage condition
    (4) Wire—open circuit or short circuit.

It is emphasized again that this last list constitutes failure mechanisms for the systems man and failure modes for the subsystems man. It is also a list of failure effects from the standpoint of the component designer. Let us try to imagine what sort of list the component designer would make. See Table V-2.

Table V-2. Doorbell Failure Analysis

Failure Effect	Failure Mode	Mechanism
Switch fails to make contact	• Contacts broken • High contact resistance	• Mechanical shock • Corrosion
Bell-solenoid unit fails to ring	• Clapper broken or not attached • Clapper stuck • Solenoid link broken or stuck • Insufficient magneto-motive force	• Shock • Corrosion • Open circuit in solenoid • Short circuit in solenoid
Low voltage from battery	• No electrolyte • Positive pole broken	• Leak in casing • Shock

From the table we see that the system failure modes constitute the various types of system failure. In fault tree terminology these are the "top events" that the system analyst can consider. He will select one of these top events and investigate the immediate causes for its occurrence. These immediate causes will be the immediate failure mechanisms for the particular system failure chosen, and will constitute failures of certain subsystems. These latter failures will be failure modes for the subsystems man and will make up the second level of our fault tree. We proceed, step by step, in this "immediate cause" manner until we reach the component failures. These components are the basic causes defined by the limit of resolution of our tree.

If we consider things from the component designer's point of view, all of the subsystem and system failures higher in the tree represent failure effects—that is, they represent the results of particular component failures. The component designer's failure modes would be the component failures themselves. If the component designer were to construct a fault tree, any one of these component failures could constitute a suitable top event. In other words, the designer's "system" is the component itself. The lower levels of the designer's fault tree would consist of the mechanisms or causes for the component failure. These would include quality control effects, environmental effects, etc., and in many cases would be beneath the limit of resolution of the system man's fault tree.

6. The "Immediate Cause" Concept

Now getting back to the system analyst, he first, as we have seen, defines his system (i.e., determines, its boundary) and then selects a particular system failure mode for further analysis. The latter constitutes the top event of the system analyst's fault tree. He next determines the immediate, necessary, and sufficient causes for the occurrence of this top event. It should be noted that these are not the basic causes of the event but the immediate causes or immediate mechanisms for the event. This is an extremely important point which will be clarified and illustrated in later examples.

The immediate, necessary, and sufficient causes of the top event are now treated as sub-top events and we proceed to determine their immediate, necessary, and sufficient causes. In so doing, we place ourselves in the position of the subsystems man for whom our failure mechanisms are the failure modes; that is, our sub-top events correspond to the top events in the subsystem man's fault tree.

In this way we proceed down the tree continually transferring our point of view from mechanism to mode, and continually approaching finer resolution in our mechanisms and modes, until ultimately, we reach the limit of resolution of our tree. This limit consists of basic component failures of one sort or another. Our tree is now complete.

As an example of the application of the "immediate cause" concept, consider the simple system in Figure V-2.

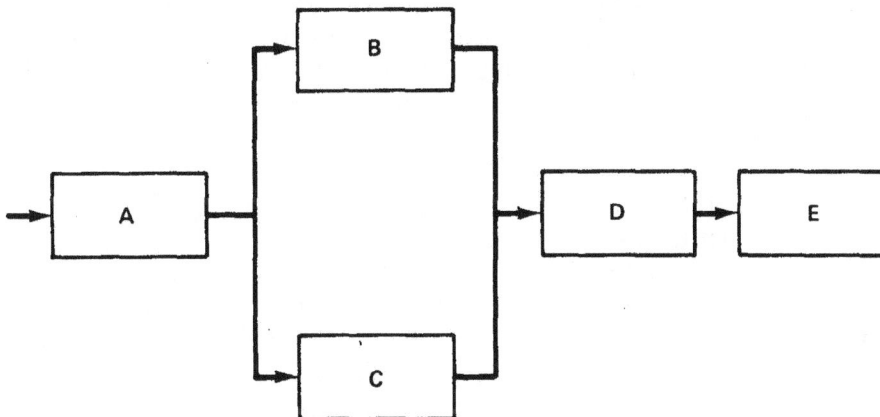

Figure V-2. System Illustrating "Immediate Cause" Concept

This system is supposed to operate in the following way: a signal to A triggers an output from A which provides inputs to B and C. B and C then pass a signal to D which finally passes a signal to E. A, B, C and D are dynamic subsystems. Furthermore, subsystem D needs an input signal from either B or C or both to trigger its output to E. We thus have redundancy in this portion of our system.

The system of Figure V-2 can be interpreted quite generally. For example, it could represent an electrical system in which the subsystems are analog modules (e.g., comparators, amplifiers, etc.); it could be a piping system in which A, B, C and D are valves; or it could represent a portion of the "chain of command" in a corporation.

Let us choose for our top event the possible outcome "no signal to E," and let us agree that in our analysis we shall neglect the transmitting devices (passive components) which pass the signals from one subsystem to another. This is tantamount to assigning a zero failure probability to the wires, pipes, or command links.

We now proceed to a step-by-step analysis of the top event. The immediate cause of the event, "no signal to E," is "no output from D." The analyst should strongly resist the temptation to list the event, "no input to D" as the immediate cause of "no signal to E." In the determination of immediate causes, one step should be taken at a time. The "immediate cause" concept is sometimes called the "Think Small" Rule because of the methodical, one-step-at-a-time approach.

We have now identified the sub-top event, "no output from D," and it is next necessary to determine its immediate cause or causes. There are two possibilities:

(1) "There is an input to D but no output from D."
(2) "There is no input to D."

Therefore, our sub-top event, "no output from D," can arise from the union of the two events, 1 or 2. The reader should note that if we had taken more than one step and had identified (improperly) the cause of "no input to D," then event 1 above would have been missed. In fact, the motivation for considering immediate causes is now clear: it provides assurance that no fault event in the sequence is overlooked.

We are now ready to seek out the immediate causes for our new mode failures, events 1 and 2. If our limit of resolution is the subsystem level, then event 1 (which can be rephrased, "D fails to perform its proper function due to some fault internal to D") is not analyzed further and constitutes a basic input to our tree. With respect to event 2, its immediate, necessary and sufficient cause is "no output from B and no output from C," which appears as an intersection of two events, i.e.,

$$2 = 3 \text{ and } 4$$

where

$3 = $ "no output from B," and
$4 = $ "no output from C."

As a matter of terminology, it is convenient to refer to events as "faults" if they are analyzed further (e.g., event 2). However, an event such as 1 which represents a basic tree input and is not analyzed further is referred to as a "failure." This terminology is also fairly consistent with the mechanistic definitions of "fault" and "failure" given previously.

We must now continue the analysis by focusing our attention on events 3 and 4. As far as 3 is concerned, we have

$$3 = 5 \text{ or } 6$$

where

$5 = $ "input to B but no output from B," and
$6 = $ "no input to B."

We readily identify 5 as a failure (basic tree input). Event 6 is a fault which can be analyzed further. We deal with event 4 in an analogous way.

The further steps in the analysis of this system can now be easily supplied by the reader. The analysis will be terminated when all the relevant basic tree inputs have been identified. In this connection, the event "no input to A" is also considered to be a basic tree input.

Our analysis of the top event ("no input to E") consequently produced a linkage of fault events connected by "and" and "or" logic. The framework (or system model) on which this linkage is "hung" is the fault tree. The next section provides the necessary details for connecting the fault event linkage to its framework (fault tree).

At this juncture the reader may be interested in conducting a short analysis of his own with reference to the doorbell circuit previously given in Figure V-1. A suitable top event would be "doorbell fails to ring when finger pushes button." The analysis would commence with the statement:

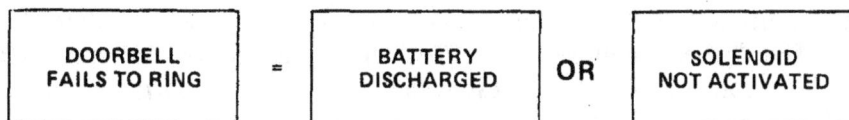

DOORBELL FAILS TO RING	=	BATTERY DISCHARGED	OR	SOLENOID NOT ACTIVATED

in which the event "battery discharged" represents a failure or basic tree input.

7.   Basic Rules for Fault Tree Construction

The construction of fault trees is a process that has evolved gradually over a period of about 15 years. In the beginning it was thought of as an art, but is was soon realized that successful trees were all drawn in accordance with a set of basic rules. Observance of these rules helps to ensure successful fault trees so that the process is now less of an art and more of a science. We shall now examine the basic rules for successful fault tree analysis.

Consider Figure V-3. It is a simple fault tree or perhaps a part of a larger fault tree. Note that none of the failure events have been "written in"; they have been designated just Q, A, B, C, D.

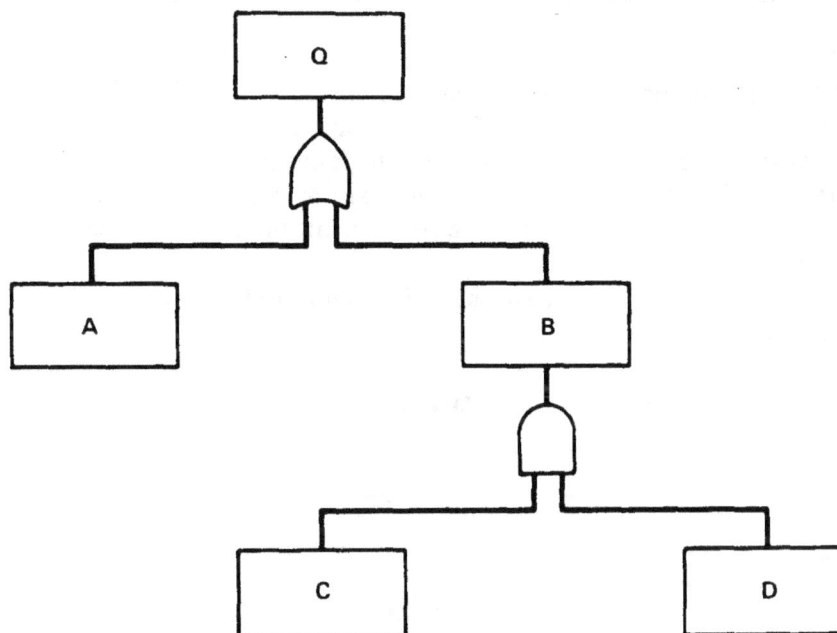

Figure V-3. A Simple Fault Tree

Now, when we have a specific problem in hand, it becomes necessary to describe exactly what such events such as Q, A, B, C, D actually are, and the proper procedure for doing this constitutes **Ground Rule I**:

> **Write the statements that are entered in the event boxes as faults; state precisely what the fault is and when it occurs.**

The "what-condition" describes the relevant railed (or operating) state of the component. The "when-condition" describes the condition of the system—with respect to the component of interest—which makes that particular state of existence of the component a fault.

Note that Ground Rule I may frequently require a fairly verbose statement. So be it. The analyst is cautioned not to be afraid of wordy statements. Do not tailor the length of your statement to the size of the box that you have drawn. If necessary, make the box bigger! It is permissible to abbreviate words but resist the temptation to abbreviate ideas. Examples of fault statements are:

(1) Normally closed relay contacts fail to open when EMF is applied to coil.

(2) Motor fails to start when power is applied.

The next step in the procedure is to examine each boxed statement and ask the question: "Can this fault consist of a component failure?" This question and its answer leads us to **Ground Rule II**:

> **If the answer to the question, "Can this fault consist of a component failure?" is "Yes," classify the event as a "state-of-component fault." If the answer is "No," classify the event as a "state-of-system fault."**

If the fault event is classified as "state-of-component," add an OR-gate below the event and look for primary, secondary and command modes. If the fault event is classified as "state-of-system," look for the minimum necessary and sufficient immediate cause or causes. A "state-of-system" fault event may require an AND-gate, an OR-gate, an INHIBIT-gate, or possibly no gate at all. As a general rule, when energy originates from a point outside the component, the event may be classified as "state-of-system."

To illustrate Ground Rule II, consider the simple motor-switch-battery circuit depicted in Figure V-4.

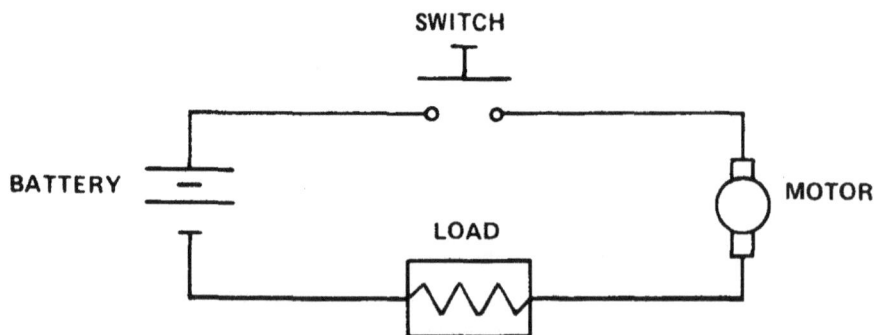

Figure V-4. Simple Motor-Switch-Battery System

The system can exist in two states: operating and standby. The following faults can be identified and classified using Ground Rule II:

OPERATING STATE

SYSTEM FAULT	CLASSIFICATION
Switch fails to close when thumb pressure is applied.	State of component
Switch inadvertently opens when thumb pressure is applied.	State of component
Motor fails to start when power is applied to its terminals.	State of component
Motor ceases to run with power applied to terminals.	State of component

STANDBY STATE

SYSTEM FAULT	CLASSIFICATION
Switch inadvertently closes with no thumb pressure applied.	State of component
Motor inadvertently starts.	State of system

In addition to the above ground rules, there are a number of other procedural statements that have been developed over the years. The first of these is the **No Miracles Rule:**

If the normal functioning of a component propagates a fault sequence, then it is assumed that the component functions normally.

We might find, in the course of a system analysis, that the propagation of a particular fault sequence could be blocked by the miraculous and totally unexpected <u>failure</u> of some component. The correct assumption to make is that the component functions normally, thus allowing the passage of the fault sequence in question. However, if the normal functioning of a component acts to block the propagation of a fault sequence, then that normal functioning must be defeated by faults if the fault sequence is to continue up the tree. Another way of stating this is to say that, if an AND situation exists in the system, the model must take it into account.

Two other procedural statements address the dangers of not being methodical and attempting to shortcut the analysis process. The first is the **Complete-the-Gate Rule:**

All inputs to a particular gate should be completely defined before further analysis of any one of them is undertaken.

The second is the **No Gate-to-Gate Rule**:

> **Gate inputs should be properly defined fault events, and gates should not be directly connected to other gates.**

The Complete-the-Gate Rule states that the fault tree should be developed in levels, and each level should be completed before any consideration is given to a lower level. With regard to the No-Gate-to-Gate Rule, a "shortcut" fault tree is shown in Figure V-5.

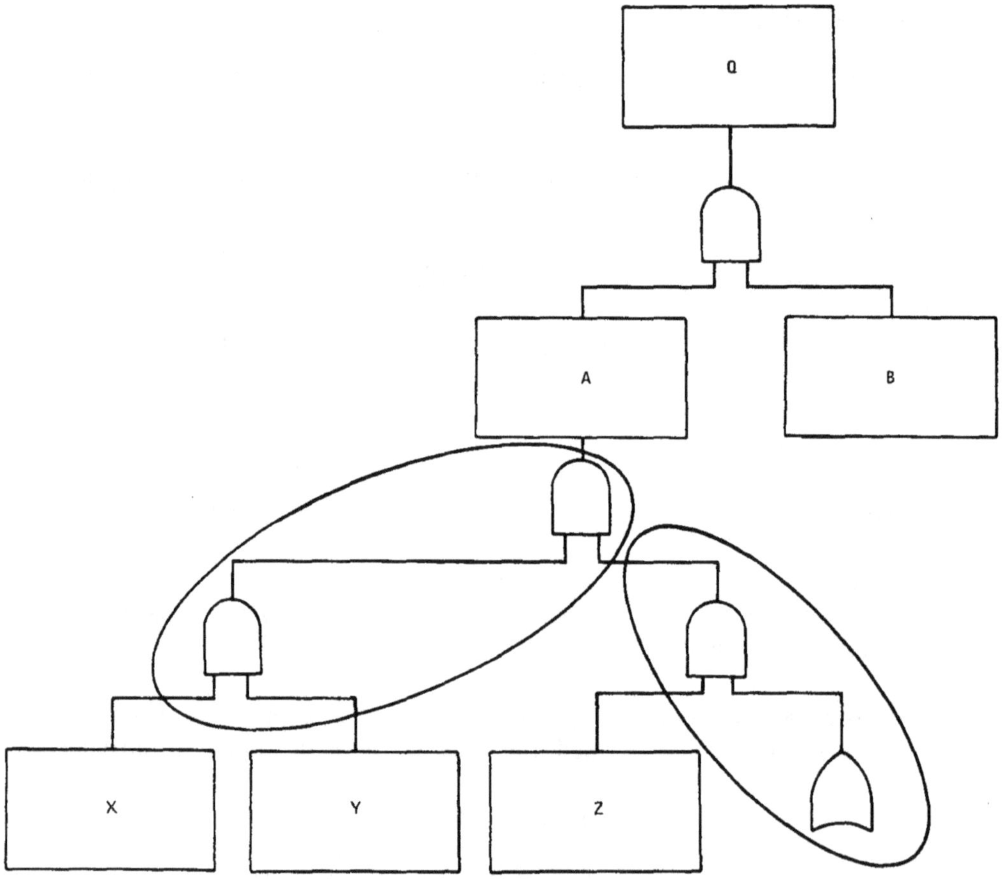

Figure V-5. A Short-Cut Fault Tree

The "gate-to-gate" connection is indicative of sloppy analysis. The "gate-to-gate" shortcutting may be all right if a quantitative evaluation is being performed and the fault tree is being summarized. However, when the tree is actually being constructed, the gate-to-gate shortcuts may lead to confusion and may demonstrate that the analyst has an incomplete understanding of the system. A fault tree can be successful only if the analyst has a clear and complete understanding of the system to be modeled.

CHAPTER VI – PROBABILITY THEORY: THE MATHEMATICAL DESCRIPTION OF EVENTS

1. Introduction

Having completed our discussion of fault tree fundamentals, we are almost ready to begin some actual fault tree construction examples. However, because the examples we will be discussing in Chapters VIII and IX include not only construction but evaluation of the trees, we must first digress in Chapters VI and VII to cover some basic mathematical concepts which underlie that evaluation.

Chapter VI addresses the basic mathematical technique involved in the quantitative assessment of fault trees: probability theory. Probability theory is basic to fault tree analysis because it provides an analytical treatment of <u>events</u>, and events are the fundamental components of fault trees. The topics of probability theory which we shall consider include the concepts of outcome collections and relative frequencies, the algebra of probabilities, combinatorial analysis, and some set theory. We begin with the concept of outcome collections which can be conveniently described in terms of a random experiment and its outcomes.

2. Random Experiments and Outcomes of Random Experiments

A random experiment is defined to be any observation or series of observations in which the possible result or results are non-deterministic. A result is deterministic if it always occurs as the outcome of an observation; a result is non-deterministic if it is only one of a number of possibilities that may occur. Thus, if we toss an ordinary coin and our purpose is to determine whether the coin will land Heads or Tails, we are performing a random experiment, which is the flipping of the coin. If our coin is "phony," however, and possesses, say, two Tails, we are not performing a random experiment in the sense of the definition because the outcome, Tails, may be expected on every toss. Similarly, the throw of a die constitutes a random experiment unless the die is loaded to give exactly the same outcome on every trial. The term "random experiment" is a very general one and the reader will be able to supply innumerable examples, such as: the measurement of stiffness of a certain spring, the time to failure of a motor, the measurement of the amount of iron in a meteorite, etc. It is clear that many non-trivial observations of our environment constitute random experiments.

A random experiment may be characterized by itemizing all its possible outcomes. This is easily done in simple cases when the number of outcomes is not great, but can also be done, at least in theory, when the number of outcomes is very great, perhaps infinite. The itemization of the outcomes of a random experiment is known, mathematically, as an <u>outcome space</u> but we believe that the term <u>outcome collection</u> is somewhat more descriptive. The notation $\{E_1, E_2, \ldots, E_n\}$ will be used to denote the outcome collection of possible events $E_1, E_2, \ldots, E_n$. The concept of an outcome collection will now be illustrated by means of several simple examples.

(a) Random experiment—one toss of a fair coin.
     Object—to determine whether coin lands Heads (H) or Tails (T).
     Outcome collection: {T, H}.
   Note that if there is present danger of the coin's disappearing into the depths of a
nearby crevasse, then this latter outcome (or event) should be added to the outcome
collection.

(b) Random experiment—toss of a coin onto a ruled surface.
     Object—to determine the coordinates of the coin when it has come to rest.
     Outcome collection: $\{(x_1, y_1), (x_2, y_2), (x_3, y_3), \ldots\}$ where (x_1, y_1) are the
Cartesian coordinates which locate the coin on the ruled surface. Note that this
outcome collection has an infinite number of elements.

(c) Random experiment—start of a diesel.
     Object—to determine whether the diesel starts (S) or fails to start (F).
     Outcome collection: {S, F}.

(d) Random experiment—the throw of a die.
     Object—to determine what number is uppermost when the die comes to rest.
     Outcome collection: {1, 2, 3, 4, 5, 6}.

(e) Random experiment—the operation of a diesel which has successfully started.
     Object—to determine the time of failure (t) to the nearest hour.
     Outcome collection: $\{t_1, t_2, \ldots\}$.

(f) Random experiment—an attempt to close a valve.
     Object—to determine if valve closes (C) or remains open (O).
     Outcome collection: {C, O}.
   If, in addition, we were to consider partial failure modes, the outcome collection
would contain such further events as "Valve closes less than halfway," "Valve closes,
then opens," etc.

(g) Random experiment—the operation of a system for some prescribed length of
time t. The system includes two critical components, A and B, and will fail if either
or both of these fail. (A and B are thus single failures.)
     Object—To determine if (and how) the system fails by time t.
     Outcome collection: {System does not fail,
                          System fails due to failure of A alone,
                          System fails due to failure of B alone,
                          System fails due to joint failure of A and B}.
   Note that the outcome collection does not include the time of failure because we
are observing only whether the system fails or succeeds in some given time.

(k) Random experiment—operation of two parallel systems. If the two systems are
designated A and B, respectively, we can define F_i = the event that system i fails and
O_i = the event that system i does not fail (i = A, B).
     Object—to determine the state of the two systems after a given time interval.
     Outcome collection: $\{(F_A, O_B), (O_A, F_B), (O_A, O_B), (F_A, F_B)\}$.
   In this example only the event (F_A, F_B) constitutes overall system failure.

3. The Relative Frequency Definition of Probability

Consider some general random experiment with the outcome collection, E_1, E_2, E_3, $\ldots, E_n$. Suppose we repeat the experiment N times and watch for the occurrence of some specific outcome, say, E_1. After N repetitions or trials we find, by actual count, that outcome E_1 has happened N_1 times. We then set up the ratio,

$$\frac{N_1}{N}$$

which represents the relative frequency of occurrence of E_1 in exactly N repetitions of this particular random experiment. The question now arises: does this ratio approach some definite limit as N becomes very large $(N \to \infty)$? If it does, we call the limit the probability associated with event E_1, in symbols, $P(E_1)$. Thus

$$P(E_1) = \lim_{N \to \infty} \left(\frac{N_1}{N}\right) \qquad \text{(VI-1)}$$

Some obvious properties of $P(E_1)$ arise from this definition:

$0 < P(E_1) < 1$
If $P(E_1) = 1$, E_1 is certain to occur.
If $P(E_1) = 0$, E_1 is impossible.

A more formal definition of probability involves set theory; however, the "empirical" definition given by Equation (VI-1) is adequate here.

4. Algebraic Operations with Probabilities

Consider a random experiment and designate two of its possible outcomes as A and B. Suppose that A and B are mutually exclusive. This simply means that A and B cannot both happen on a single trial of the experiment. For instance, we expect to get either Heads or Tails as a result of a coin toss. We could not possibly get both Heads and Tails on a single toss. If events A and B are mutually exclusive, we can write down an expression for the probability that either A or B occurs:

$$P(A \text{ or } B) = P(A) + P(B) \qquad \text{(VI-2)}$$

This relation is sometimes referred to as the "addition rule for probabilities" and is applicable to events that are mutually exclusive. This formula can be readily extended to any number of mutually exclusive events A, B, C, D, E, ...

$$P(A \text{ or } B \text{ or } C \text{ or } D \text{ or } E) = P(A) + P(B) + P(C) + P(D) + P(E) \qquad \text{(VI-3)}$$

For events which are not mutually exclusive a more general formula must be used. For example, suppose that the random experiment is the toss of a single die and let us define two events as follows:

A—"the number 2 turns up"
B—"an even number turns up"

Clearly these events are not mutually exclusive because if the result of the toss is 2, both A and B have occurred. The general expression for P(A or B) is now

$$P(A \text{ or } B) = P(A) + P(B) - P(A \text{ and } B) \qquad \text{(VI-4)}$$

If A and B are mutually exclusive, P(A and B) = 0 and (VI-4) reduces to (VI-2). The reader should also note that (VI-2) always gives an upper bound to the true probability (VI-4) when the events are not mutually exclusive. Now, if we return to our single die problem and define A and B as above, we can calculate P(A or B) numerically:

$$P(A \text{ or } B) = 1/6 + 1/2 - 1/6 = 1/2$$

Equation (IV-4) can be extended to any number of events. For example, for 3 events A, B, C

$$P(A \text{ or } B \text{ or } C) = P(A) + P(B) + P(C) - P(A \text{ and } B) - P(A \text{ and } C)$$

$$- P(B \text{ and } C) + P(A \text{ and } B \text{ and } C) \qquad \text{(VI-5)}$$

For n events $E_1, E_2, \ldots, E_n$, the general formula can be expressed as:

$$P(E_1 \text{ or } E_2 \text{ or } \ldots \text{ or } E_n) = \sum_{i=1}^{n} P(E_i) - \sum_{i=1}^{n-1} \sum_{j=i+1}^{n} P(E_i \text{ and } E_j)$$

$$+ \sum_{i=1}^{n-2} \sum_{j=i+1}^{n-1} \sum_{k=j+1}^{n} P(E_i \text{ and } E_j \text{ and } E_k) \ldots$$

$$+ (-1)^n P(E_1 \text{ and } E_2 \text{ and } \ldots \text{ and } E_n) \qquad \text{(VI-6)}$$

where "Σ" is the summation sign.

If we ignore the possibility of any two or more of the events E_i occurring simultaneously, equation (VI-6) reduces to

$$P(E_1 \text{ or } E_2 \text{ or } \ldots \text{ or } E_n) = \sum_{i=1}^{n} P(E_i) \qquad \text{(VI-7)}$$

Equation (VI-7) is the so-called "rare event approximation" and is accurate to within about ten percent of the true probability when $P(E_i) < 0.1$. Furthermore, any error made is on the conservative side, in that the true probability is slightly lower than that given by equation (VI-7). The rare event approximation plays an important role in fault tree quantification and is discussed further in Chapter XI.

Consider now two events A and B that are <u>mutually independent</u>. This means that in the course of several repetitions of the experiment, the occurrence (or non-occurrence) of A has no influence on the subsequent occurrence (or non-occurrence) of B and vice versa. If a well-balanced coin is tossed randomly, the occurrence of Heads on the first toss should not cause the probability of Tails on the second toss to be any different from 1/2. So the results of successive tosses of a coin are considered to be mutually independent outcomes. Also, if two components are operating in parallel and are isolated from one another, then the failure of one does not affect the failure of the other. In this case the failures of the components are independent events. If A and B are two mutually independent events, then we can write,

$$P(A \text{ and } B) = P(A) P(B) \tag{VI-8}$$

This is often called the "multiplication rule for probabilities" and its extension to more than two events is obvious.

$$P(A \text{ and } B \text{ and } C \text{ and } D) = P(A) P(B) P(C) P(D) \tag{VI-9}$$

Very often, we encounter events that are <u>not</u> mutually independent, that is, they are mutually <u>inter</u>dependent. For instance, the overheating of a resistor in an electronic circuit may very well change the failure probability of a nearby transistor or of related circuitry. The probability of rain on Tuesday will most likely be influenced by the weather conditions prevailing on Monday. In order to treat events of this nature, we introduce the concept of conditional probability, and we need a new symbol: $P(B|A)$ which is the probability of B, given A has already occurred. The probability of A and B both occurring then becomes

$$P(A \text{ and } B) = P(A) P(B|A) = P(B) P(A|B). \tag{VI-10}$$

If A and B are mutually independent, then $P(A|B) = P(A)$ and $P(B|A) = P(B)$ and (VI-10) reduces to (VI-8). Thus Equation (VI-10) constitutes a general expression for the probability of the joint occurrence of two events A and B.

For three events A, B, C, we have

$$P(A \text{ and } B \text{ and } C) = P(A) P(B|A) P(C|A \text{ and } B) \tag{VI-11}$$

where $P(C|A \text{ and } B)$ is the probability of C occurring given A and B have already occurred. For n events $E_1, E_2, \ldots E_n$,

$$P(E_1 \text{ and } E_2 \text{ and } \ldots E_n) = P(E_1) P(E_2|E_1) P(E_3|E_1 \text{ and } E_2)$$

$$\ldots P(E_n|E_1 \text{ and } E_2 \ldots \text{ and } E_{n-1}) \tag{VI-12}$$

An example of the use of the formulae in this section may be helpful. Consider the following random experiment: we select one card at random from a well-shuffled regulation deck of 52. We note its face value (e.g., "seven," "Queen," etc.) and lay it aside (i.e., we do not put it back in the deck—this is known as "sampling without replacement"). We then choose a second card at random from the deck of 51 and note its face value. We now calculate the probability of getting at least one Ace in the two draws. There are three mutually exclusive possibilities:

(Ace on first draw then non-Ace on second draw) or
(non-Ace on first draw then Ace on second draw) or
(Ace on first draw then Ace again on second draw).

Expressed in mathematical language, this becomes much more succinct:

P (at least one Ace in two draws) = P(A)

$$= P(A_1 \text{ and } \overline{A}_2) + P(\overline{A}_1 \text{ and } A_2) + P(A_1 \text{ and } A_2)$$

$$= P(A_1) P(\overline{A}_2 | A_1) + P(\overline{A}_1) P(A_2 | \overline{A}_1) + P(A_1) P(A_2 | A_1)$$

in which the subscripts refer to the order of the draw, A stands for "Ace" and $\overline{A}$ stands for "non-Ace." We can evaluate this expression numerically as follows:

$$P(A) = \left(\frac{4}{52}\right)\left(\frac{48}{51}\right) + \left(\frac{48}{52}\right)\left(\frac{4}{51}\right) + \left(\frac{4}{52}\right)\left(\frac{3}{51}\right) = \frac{396}{(52)(51)} = \frac{33}{221} = 0.149$$

Let us now calculate the probability of getting an Ace and a King in two draws.

$$P(A \text{ and } K) = P(A_1) P(K_2 | A_1) + P(K_1) P(A_2 | K_1)$$

$$= \left(\frac{4}{52}\right)\left(\frac{4}{51}\right) + \left(\frac{4}{52}\right)\left(\frac{4}{51}\right) = \frac{32}{(52)(51)} = \frac{8}{663} = 0.012$$

We now note an important point. If the events "getting an Ace" and "getting a King" were independent, we would have P(A and K) = P(A) P(K) = $(0.149)^2$ = 0.022, but we have just shown that P(A and K) = 0.012 ≠ 0.022. Therefore, the events in question are not independent. At this point the reader is invited to present an argument that the events in question would have been independent if the first card drawn had been replaced in the deck and the deck had been shuffled before the second draw.

A result with important applications to fault tree analysis is the calculation of the probability of occurrence of at least one of a set of mutually independent events.

Consider the set of mutually independent events,

$$\{E_1, E_2, E_3, \ldots, E_n\}$$

and define the event $\overline{E}_1$ as the non-occurrence of E_1, the event $\overline{E}_2$ as the non-occurrence of E_2, etc. Because a particular event must either occur or not occur, we have

$$P(E_i) + P(\overline{E}_i) = 1 \qquad\qquad (VI\text{-}13)$$

$$P(\overline{E}_i) = 1 - P(E_i) \qquad\qquad (VI\text{-}14)$$

Now there are two possibilities with regard to the events $E_1, E_2, \ldots, E_n$: at least one E_i occurs or none of the E_i occur. Therefore,

P (at least one E_i occurs)

$$= 1 - P \text{ (no } E_i \text{ occurs)}$$

$$= 1 - P(\overline{E}_1 \text{ and } \overline{E}_2 \ldots \text{ and } \overline{E}_n)$$

We know that the E's are mutually independent and, for this reason, it is perhaps intuitively obvious that $\overline{E}$'s must also be mutually independent. As a matter of fact, this result can be easily proven. Therefore,

$$P(\overline{E}_1 \text{ and } \overline{E}_2 \text{ and } \overline{E}_3 \ldots \text{ and } \overline{E}_n) = P(\overline{E}_1)\, P(\overline{E}_2)\, P(\overline{E}_3) \ldots P(\overline{E}_n) \qquad (VI\text{-}15)$$

But from Equation (VI-14) this is the same as

$$\left[1 - P(E_1)\right] \left[1 - P(E_2)\right] \ldots \left[1 - P(E_n)\right] \qquad\qquad (VI\text{-}16)$$

so that our final result is

$$P(E_1 \text{ or } E_2 \text{ or } E_3 \text{ or } \ldots \text{ or } E_n) = 1 - \left\{ \left[1 - P(E_1)\right] \left[1 - P(E_2)\right] \right. \\ \left. \left[1 - P(E_3)\right] \ldots \left[1 - P(E_n)\right] \right\} \quad (VI\text{-}17)$$

In the especially simple case where $P(E_1) = P(E_2) = \ldots = P(E_n) = p$, the right-hand side of (VI-17) reduced to $1 - (1 - p)^n$.

Our general result (VI-17) finds application in fault tree evaluation. For example, consider a system in which system failure occurs if any one of the events E_1, $E_2, \ldots, E_n$ occurs, it being assumed that these events are mutually independent. The probability of system failure is then given by (VI-17). For example, the events $E_1, \ldots, E_n$ may be failures of critical components of the system, where each component failure will cause system failure. If the component failures are independent, then Equation (VI-17) is applicable. In the general case, the events E_1, E_2, etc., represent the modes by which system failure (top event of the fault tree) can occur. These modes are termed the minimal cut sets of the fault tree and if they are independent, i.e., no minimal cut sets have common component failures, then Equation (VI-17) applies. We shall discuss minimal cut sets in considerable detail in later chapters.

To conclude this section we return to the concept of an outcome collection because we are now in a position to define it in more detail. The elements of an outcome collection possess several important characteristics:

(a) The elements of an outcome collection are all <u>mutually exclusive</u>.

(b) The elements of an outcome collection are collectively exhaustive. This means that we have included in the outcome collection every conceivable result of the experiment.

(c) The elements of the outcome space may be characterized as being continuous or discrete; e.g., the times-to-failure for some systems are continuous events and the 52 possible cards when a card is drawn are discrete events.

5.   Combinatorial Analysis

We now discuss combinatorial analysis because it allows us to easily evaluate the probabilities of various combinations of events, such as failures of redundant systems. As an introduction, we review the difference between "combination" and "permutation." Consider a collection or set of four entities: {A, B, C, D}. Suppose we make a random choice of three items from the four available. For example, we may end up with the elements ABD. This group of three may be rearranged or permuted in six different ways: ABD, ADB, BAD, BDA, DAB, DBA. Thus, there are six rearrangements or permutations corresponding to the single combination ABD.

Briefly, when we talk about permutations, we are concerned with order; when we talk about combinations, we are not concerned with order. Whether order is important to us or not will depend on the specific nature of the problem. For a certain redundant system to fail we may need only some particular number of component failures without regard to the order in which they occur. For example, in a two-out-of-three logic system we may need any two-out-of-three component failures to induce system failure. In this case we are concerned with combinations of events. However, when a particular sequence of events must occur, we are generally concerned with permutations. For example, failure of containment spray injection necessarily implies failure of containment spray recirculation and hence here we are concerned with order, i.e., injection, then recirculation.

We now seek a few simple rules that will enable us to count numbers of combinations or permutations as desired. Consider the problem of choosing (at random) a sample of size r from a population of size n. This process can be conducted in two ways: (a) with replacement, and (b) without replacement. If we sample with replacement, we put each element drawn back into the population after recording its characteristics of interest. If we sample without replacement, we lay each item aside after "measuring" it. The reader will recall that this concept was mentioned briefly in our example involving the draw of two cards from a desk.

If we sample with replacement, how many different samples of size r are possible? We have n choices for the first item, n choices for the second item, n choices for the third item, and so on until a sample of size r is completed. Thus, under a replacement policy there are n^r possible samples of size r. Observe that under a replacement policy duplication of items in the sample is possible.

If we sample without replacement, we have n choices for the first item, $(n-1)$ choices for the second item, $(n-2)$ choices for the third item, $(n-r+1)$ choices for the rth item. Thus, the total number of different samples of size r that can be drawn from a population of size n without replacement is

$$(n)(n-1)(n-2)\ldots(n-r+1) \equiv (n)_r \qquad \text{(VI-18)}$$

As shown in Equation (VI-18), the special symbol $(n)_r$ is used for this product. The symbol $(n)_r$ may be written in a more familiar way by using the following algebraic device:

$$(n)_r = \frac{(n)(n-1)(n-2)\ldots(n-r+1)(n-r)(n-r-1)\ldots 3\cdot 2\cdot 1}{(n-r)(n-r-1)\ldots 3\cdot 2\cdot 1}$$

$$= \frac{n!}{(n-r)!}$$

(VI-19)

using the definition of the <u>factorial numbers</u>.

Equation (VI-19) constitutes one of the fundamental relationships that we have been seeking. It gives us <u>the number of permutations of n things taken r at a time.</u> When $r = n$, we get $(n)_n = \frac{n!}{0!} = n!$ because $0! \equiv 1$. Thus the number of ways of rearranging n items among themselves is just $n!$.

If we are interested in combinations and not permutations, we do not desire to count the $r!$ ways in which the r items of the sample can be rearranged among themselves. Thus the number of combinations of n things taken r at a time is given by

$$\frac{n!}{(n-r)!\,r!} \equiv \binom{n}{r}$$

(VI-20)

where $\binom{n}{r}$ is the symbol used to denote this quantity. If we choose at random three elements from the population $\{A, B, C, D\}$, we have $\frac{4!}{1!\,3!} = 4$ possible combinations and they are ABC, ABD, ACD, and BCD. Each of these can be permuted in 6 ways so that there should be a total of 24 permutations and this agrees with

$$\frac{n!}{(n-r)!} = \frac{4!}{1!} = 4 \times 3 \times 2 \times 1 = 24$$

Consider now a population of n items, p of which are all alike, q of which are all alike, and r of which are all alike, with $(p+q+r) = n$. For example, we might have the quantities p resistors from one manufacturer, q from the second manufacturer, and r from a third. We can distinguish among manufacturers, but resistors from the same manufacturer are treated as being indistinguishable. Thus, there will be some permutations among the $n!$ possible rearrangements that cannot be differentiated. Specifically, rearrangements of the Manufacturer 1 resistors among themselves will lead to no new distinguishable arrangements and similarly for Manufacturer 2 and Manufacturer 3. Thus, the total number of distinct arrangements is given by

$$\frac{n!}{p!\,q!\,r!}$$

(VI-21)

In general, for n items, n_1 of which are alike, n_2 of which are alike, . . . , and n_k of which are alike $(n_1 + n_2 + \ldots + n_k = n)$, the number of distinct arrangements is given by

$$\frac{n!}{n_1! \, n_2! \dots n_k!} \qquad \qquad \text{(VI-22)}$$

An example will show how these expressions are used in the solution of a problem. Let us try to calculate the probability of being dealt a full house (3-of-a-kind and a pair) in ordinary 5-card poker.

The total number of possible 5-card poker hands is simply the number of combinations of 52 things taken 5 at a time or

$$\binom{52}{5} = \frac{52!}{47! \, 5!} = N_{PH}$$

where the subscript PH stands for "poker hands." If we can find out how many of these contain a full house N_{FH}, then the ratio $\dfrac{N_{FH}}{N_{PH}}$ will give us the probability of being dealt a full house. First let us determine the number N_{FH}. We can write a full house symbolically as XXXYY, where X and Y stand for any two of the 13-card categories 2, 3, 4, ..., J, Q, K, A. In how many ways can the categories X, Y be chosen? Clearly we have 13 possible choices for X and then 12 possible choices for Y so that the product (13)(12) represents the total number of ways of selecting the categories X and Y. Now there are 4 suits and the triplet XXX must be constituted in some way from the 4 available X's. The number of ways of accomplishing this is $\binom{4}{3}$ = 4. Similarly, the pair YY must be constituted in some way from the 4 available Y's and the number of ways of doing this is $\binom{4}{2}$ = 6. Consequently N_{FH} = (13)(12)(4)(6) and

$$P(\text{Full House}) = \frac{N_{FH}}{N_{PH}} = \frac{(13)(12)(4)(6)}{\binom{52}{5}} = \frac{6}{4168} \text{ or approximately } \frac{1}{700}$$

For an example more pertinent to reliability analysis, consider a system consisting of n similar components; the system fails if m out of n components fail. For example, the system may consist of 3 sensors where 2 or more sensor failures are required for system failure (2-out-of-3 logic). The number of ways the system can fail if m components fail is $\binom{n}{m}$, or the number of combinations of n items taken m at a time. If any component has a probability p of failing, the probability of any one combination leading to system failure is

$$p^m \, (1-p)^{n-m}$$

Thus the probability of system failure from m components failing is

$$\binom{n}{m} p^m \, (1-p)^{n-m}$$

In addition to m components failing, the system can also fail if m+1 components fail or if m+2 components fail, etc., up to all n components failing. The number of ways the system can fail from k component failures is $\binom{n}{k}$ where k = m+1, m+2, ... , k. The probability for a particular combination of k failures is

$$p^k (1-p)^{n-k} \qquad k = m+1, m+2, \ldots, n.$$

Hence, the probability of system failure from k components failing is

$$\binom{n}{k} p^k (1-p)^{n-k} \qquad k = m+1, m+2, \ldots, n.$$

To obtain the total system failure probability we add up the probabilities for m components failing, m+1 components failing, etc., and hence the total system failure probability is

$$\binom{n}{m} p^m (1-p)^{n-m} + \binom{n}{m+1} p^{m+1} (1-p)^{n-m-1} + \ldots + \binom{n}{n} p^n (1-p)^0$$

which can also be written as

$$\sum_{k=m}^{n} \binom{n}{k} p^k (1-p)^{n-k}$$

The probabilities are examples of the binomial distribution which we shall discuss later in more detail.

6. Set Theory: Application to the Mathematical Treatment of Events

As seen in the previous section, combinatorial analysis allows us to determine the number of combinations pertinent to an event of interest. Set theory is a more general approach which allows us to "organize" the outcome events of an experiment to determine the appropriate probabilities. In the most general sense, a set is a collection of items having some recognizable features in common so that they may be distinguished from other things of different species. Examples are: prime numbers, relays, scram systems, solutions of Bessel's equation, etc. Our application of set theory involves a considerable particularization. The items of immediate interest to us are the outcome events of random experiments and our development of the elementary notions of set theory will be restricted to the event concept.

We can think of an event as a collection of elements. Consider, for example, the following possible events of interest associated with the toss of a die:

A—the number 2 turns up
B—the result is an even number
C—the result is less than 4
D—some number turns up
E—the result is divisible by 7.

Each of these events can be considered as a particular set whose elements are drawn from the basic outcome collection of the experiment: {1, 2, 3, 4, 5, 6 }.

We have:

A = {2 }
B = {2, 4, 6 }
C = {1, 2, 3 }
D = {1, 2, 3, 4, 5, 6 }
E = ϕ (the null, void, or empty set),

where braces, "{ }," are used to denote a particular set and the quantities within the braces are the elements of that set.

Event A can be represented as a set having a single element—the number 2. Both B and C can be represented as 3-element sets. Event D contains all the possible results of the experiment. It thus coincides with the outcome collection. Any such set that contains all the outcomes of an experiment is referred to as the <u>universal set</u> and is generally denoted by the symbol Ω or I (also sometimes by the number 1 when the notation is informal). E is an impossible event and can be represented by a set containing no elements at all, the so-called <u>null set</u> symbolized by ϕ.

Returning to our die-throwing example, we note that the element "1" belongs to C and D but not to A or B. This fact is symbolized as follows:

$$1 \in C, \ 1 \in D, \ 1 \notin A, \ 1 \notin B.$$

where the symbol "$\in$" means "is an element of" and the symbol $\notin$ means "is not an element of."

We also note that the elements of A, B, C are contained in set D. We call A, B, C <u>subsets</u> of D and write A⊂D, B⊂D, C⊂D. Observe also that A is a subset of both B and C. If X and Y are two sets such that X is a subset of Y, i.e., X⊂Y and Y is a subset of X, i.e., Y⊂X, then X and Y are equal (i.e., they are the same set).

As another example, consider the time of failure t of a diesel (in hours) and consider the sets

A = {t=0 }
B = {t_i, 0 < t ⩽ 1 }
C = {t_i, t > 1 }

The failure to start of the diesel is represented by A, i.e., "zero failure time." B represents times of failure which are greater than zero hours (i.e., the diesel started) and less than or equal to one hour. C represents time of failure greater than 1 hour. Each of these sets, i.e., events, could be associated with different consequences if an abnormal situation existed (i.e., loss of offsite power).

There exists a graphical procedure that permits a simple visualization of the set theoretic concept which is known as a <u>Venn diagram</u>. The universal set is represented by some <u>geometrical shape</u> (usually a rectangle) and any subsets (events) of interest are shown inside. Figure VI-1 represents a Venn diagram for the previous toss of a die example.

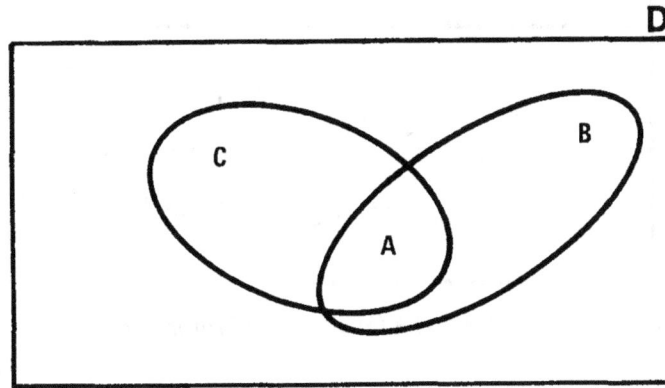

Figure VI-1. Venn Diagram Representation of Sets

Operations on sets (events) can be defined with the help of Venn diagrams. The operation of <u>union</u> is portrayed in Figure VI-2.

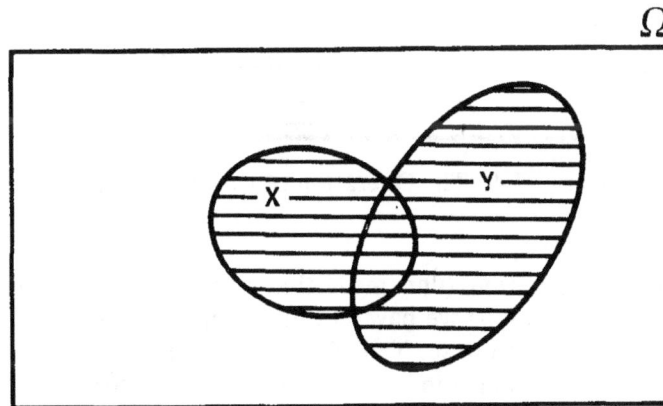

Figure VI-2. The Operation of Union

The union of two sets of X, Y is the set that contains all elements that are either in X or in Y or in both, is written as X∪Y, and is indicated by the shaded area in Figure VI-2. Returning to the die example, the union of B, C is written,

$$B \cup C = \{1, 2, 3, 4, 6\}.$$

Note that the word "or" translates into the symbol "∪."

The operation of <u>intersection</u> is portrayed in Figure VI-3. The intersection of two sets X,Y is the set that contains all elements that are common to X and Y, is written as X∩Y, and is indicated by the shaded area in Figure VI-3. In the die Example B∩C = {2} = A. Note that the word "and" translates into the symbol "∩."

The operation of <u>complementation</u> is portrayed in Figure VI-4. The complement of a set X is the set that contains all elements that are not in X, is written $\bar{X}$ (or X′), and is indicated by the shaded area in Figure VI-4. For the die example, the complement of the set B∪C is (B∪C)′ = $(\overline{B \cup C})$ = {5}.

Ω

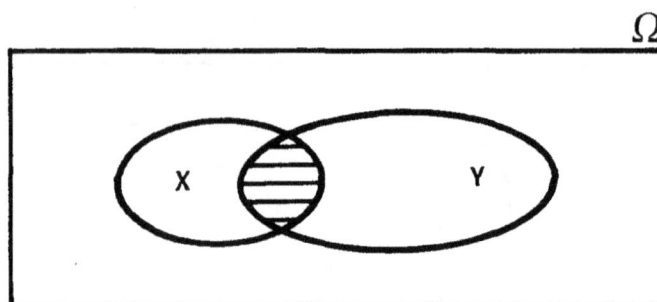

Figure VI-3. The Operation of Intersection

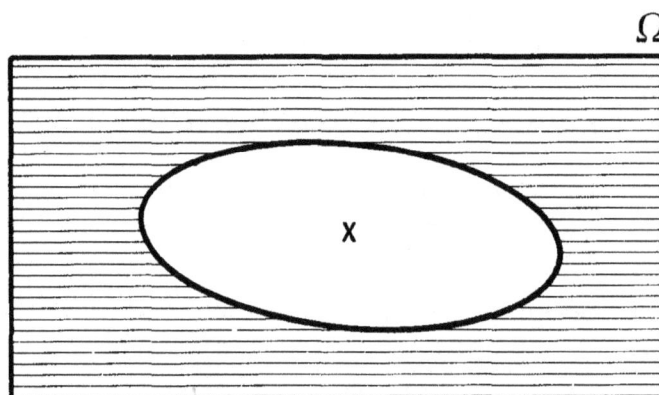

Ω

Figure VI-4. The Operation of Complementation

There is another operation (unnamed) that is sometimes defined, but it is not independent of the operations we have already given. The operation is illustrated in Figure VI-5. If we remove from set Y all elements that are common to both X and Y, we are left with the shaded area indicated in Figure VI-5. This process is occasionally written (Y−X) but the reader can readily see that

$$(Y-X) = Y \cap X'$$

Ω

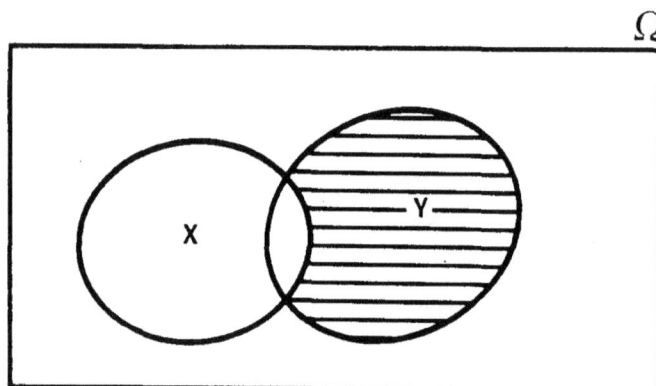

Figure VI-5. The Operation (Y−X)

so that the symbol (Y–X) is not needed and hence will not be further used.

As an example of the use of the set-theoretic approach, consider a simple system consisting of three components A, B, C. Let us use the symbols A, B, C not only to designate the components themselves but also the events "component success" for A, B, C, respectively. Thus, by $\overline{A}$, $\overline{B}$, $\overline{C}$, we shall understand the events "component failure" for A, B, C, respectively. The symbol $A\overline{B}C$ will consequently be used to represent the event, "A operates successfully, B fails, and C operates successfully."

Because there are 3 components and 2 modes of operation for each component (success/failure), combinatorial analysis tells us that there are $2^3 = 8$ combinations which give all system modes of failure or operation. Thus, the universal set (or outcome collection) is given by:

$$\Omega = \{ABC, AB\overline{C}, A\overline{B}C, A\overline{B}\overline{C}, \overline{A}BC, \overline{A}B\overline{C}, \overline{A}\overline{B}C \ \overline{A}\overline{B}C \}.$$

Suppose that we have determined (perhaps by fault tree analysis) that our system will fail if any two or more of its components fail. Then the events corresponding to system failure are:

$$\overline{S}_1 = A\overline{B}\overline{C}$$
$$\overline{S}_2 = \overline{A}B\overline{C}$$
$$\overline{S}_3 = \overline{A}\overline{B}C$$
$$\overline{S}_4 = \overline{A}\overline{B}\overline{C}$$

and the event, "system failure," S, can be represented by the subset $\overline{S} = \overline{S}_1 \cup \overline{S}_2 \cup \overline{S}_3 \cup \overline{S}_4 = \{A\overline{B}\overline{C}, \ \overline{A}B\overline{C}, \ \overline{A}\overline{B}C, \ \overline{A}\overline{B}\overline{C}\}$. We have enumerated, in an exhaustive manner, all the ways the system can fail. This information can be used in various applications. For example, knowing the probability of component failures, we can calculate the probability of system failure. In the same way we have done above, we may take intersections and unions of any of the basic elements (simple events) of the universal set to generate still other events which can be represented as sets consisting of particular elements.

This approach of enumerating the outcome events is sometimes used in the inductive methods of system analysis (according to which we enumerate all possible combinations of component operation or non-operation and determine the effects of each possibility on system behavior). The so-called "matrix approaches" are of this nature. If our system is relatively simple or if the "components" are relatively gross (e.g., subsystems), these inductive approaches can be efficient, in that the number of combinations in the universal set will be fairly small. We can also use an inductive approach to determine which combination has the most serious consequences. This latter event can then be more fully analyzed by means of fault tree analysis.

Using the set theory concepts we have developed, we can now translate probability equations into set theoretic terms. For example,

P(A or B) = P(A) + P(B) – P(A and B) becomes

$$P(A \cup B) = P(A) + P(B) - P(A \cap B). \tag{VI-23}$$

The equation $P(A \text{ and } B) = P(A|B)\ P(B) = P(B|A)\ P(A)$ becomes

$$P(A \cap B) = P(A|B)\ P(B) = P(B|A)\ P(A). \tag{VI-24}$$

We have introduced a new mathematical entity, the set (or, more particularly, the event). An algebra based on the defined operations of union, intersection, and complementation is called a Boolean algebra. By using the basic operations of union, intersection, and complementation, Boolean algebra allows us to express events in terms of other basic events. In our fault tree applications, system failure can be expressed in terms of the basic component failures by translating the fault tree to equivalent Boolean equations. We can manipulate these equations to obtain the combinations of component failures that will cause system failure (i.e., the minimal cut sets) and we calculate the probability of system failure in terms of the probabilities of component failures. We shall further pursue these topics in later sections.

7.   Symbolism

Boolean algebra, which is the algebra of events, deals with event operations which are represented by various symbols. Unfortunately, set theoretic symbolism is not uniform; the symbology differs among the fields of mathematics, logic, and engineering as follows:

Operation	Probability	Mathematics	Logic	Engineering
Union of A and B	A or B	$A \cup B$	$A \vee B$	$A + B$
Intersection of A and B	A and B	$A \cap B$	$A \wedge B$	$A \cdot B$ or AB
Complement of A	not A	A' or $\overline{A}$	$-A$	A' or $\overline{A}$

The symbols used in mathematics and in logic are very similar. The logical symbols are the older; in fact the symbol "$\vee$" is an abbreviation for the Latin word "vel" which means "or." It is unfortunate that engineering has adopted "+" for "$\cup$" and an implied multiplication for "$\cap$." This procedure "overworks" the symbols "+" and "$\cdot$". As an example of the confusion that might arise when "+" is used for $\cup$, consider the expression

$$P(A \cup B) = P(A) + P(B) - P(A \cap B).$$

If "+" is written for "$\cup$" on the left-hand side, we have an equation with "+" meaning one thing on the left and a totally different thing on the right.

Despite these difficulties and confusing elements in the symbology, the engineering symbology is now quite widespread in the engineering literature and any expectation of a return to mathematical or logic symbols seems futile. In fault tree analysis, use of the engineering notation is widespread, and, as a matter of fact, we shall use it later in this book. If, however, the reader is unacquainted with event algebra, it is strongly recommended that he use the proper mathematical symbols until attaining familiarity with this type of algebra. This will serve as a reminder that set algebraic operations are not to be confused with the operations of ordinary algebra where numbers, and not events, are manipulated.

8.   Additional Set Concepts

We shall now proceed further with certain other set concepts which will illustrate the difference between simple events and compound events. This will be useful groundwork for some of the fault tree concepts to follow and will also lead to a more rigorous definition of "probability."

Consider again the throw of a single die. The event $A = \{2\}$ is a simple event; in fact it constitutes an element of the outcome collection. In contrast, the events $B = \{2, 4, 6\}$ and $C = \{1, 2, 3\}$ are compound events. They do not constitute, per se, elements of the outcome collection, even though they are made up of elements of the outcome collection. B and C have an element in common; therefore, their intersection is non-empty (i.e., they are not "disjoint" subsets or, in probability language, they are not mutually exclusive). The elements of outcome collections are, by definition, all mutually exclusive and, thus, all mutually disjoint.

Now compound events (like B and C) are generally the ones that are of predominant interest in the real world and it is necessary, because they are not included in the outcome collection, to define a mathematical entity that does include them. Such a mathematical entity is called a class. A class is a set whose elements are themselves sets and these elements are generated by enumerating every possible combination of the members of the original outcome collection.

As an example, consider the 4-element outcome collection $S = \{1, 3, 5, 7\}$. If we list every possible combination of these 4-elements we shall generate the class $\underline{S}$ defined on the original set S as follows:

$$\underline{S} = \{1\}, \{3\}, \{5\}, \{7\}, \{1, 3\}, \{1, 5\},$$
$$\{1, 7\}, \{3, 5\}, \{3, 7\}, \{5, 7\}, \{1, 3, 5\},$$
$$\{1, 5, 7\}, \{1, 3, 7\}, \{3, 5, 7\}, \{1, 3, 5, 7\}, \{\phi\}.$$

Notice that the null set ϕ is considered an element of the class to provide a mathematical description of the "impossible event." If we count the number of elements in the class $\underline{S}$, we find 16 which is 2^4 where 4 is the number of elements in the original set S. In general, if the original set has n elements, the corresponding class* will have 2^n elements.

The utility of the class concept is simply that the class will contain as elements, every conceivable result (both simple and compound) of the experiment. Thus, in the die toss experiments, S will have 6 elements and $\underline{S}$ will have $2^6 = 64$ elements, two of which will be $B = \{2, 4, 6\}$ and $C = \{1, 2, 3\}$. In the throw of 2 dice, S will contain 36 elements and $\underline{S}$ will have 2^{36} (a number in excess of 10^{10}) elements. Embedded somewhere in this enormous number of subsets we find the compound event "sum = 7" which can be represented in the following way:

$$E = \left\{(1, 6), (2, 5), (3, 4), (4, 3), (5, 2), (6, 1)\right\}.$$

_____

*When certain appropriate mathematical restrictions are imposed, a class is often referred to, in more advanced texts, as a Borel Field.

Consider again the simple three-component system A, B, C, where system failure consists of any two or more components failing. In that example, the universal set comprised eight elements, and these eight elements gave all modes of system failure or operation. The class based on this set would contain 2^8 = 256 elements and would include such events as:

"A operates properly": $\{ABC, AB\overline{C}, A\overline{B}C, A\overline{B}\,\overline{C}\}$

"Both B and C fail": $\{AB\overline{C}, \overline{A}B\overline{C}\}$

"Five components fail": ϕ

"System fails": $\{A\overline{B}\,\overline{C}, \overline{A}B\overline{C}, \overline{A}\,\overline{B}C, \overline{A}\,\overline{B}\,\overline{C}\}$

The reader should bear in mind that the elements of classes are <u>sets</u>. Thus, the event ABC is an element of the original universal set, whereas {ABC} is a set containing the single element ABC and is an element of the class generated from the universal set. The utility of the class concept is that it enables us to treat compound events in a formal manner simply because all possible compound events are included in the class.

Perhaps the most useful feature of the class concept is that it provides us with a basis for establishing a proper mathematical definition of the probability function. The set theoretic definition of the probability function is shown schematically in Figure VI-6.

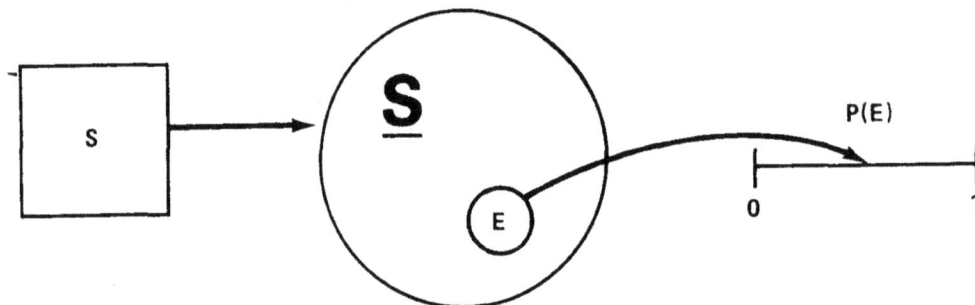

Figure VI-6. Set Theoretic Definition of Probability

In Figure VI-6, the box labeled "S" represents the outcome collection for some random experiment. It could, for instance, represent the totality of the 36 possible outcomes in the two-dice experiment. The circle labeled "<u>S</u>" represents the <u>class</u> generated from <u>set</u> S by enumerating all combinations of the elements of S. In the two-dice example the class <u>S</u> possesses an enormous number ($< 10^{10}$) of elements. These elements represent every conceivable outcome (simple and compound) of the experiment. Specifically, the event E = "sum of 7" is a member of <u>S</u>. It is shown schematically in Figure VI-6. Next the axis of real numbers between 0 and 1 is drawn. A function can now be defined that "maps" event E into some position on this axis. This function is the probability function P(E).

The concept of mapping may be unfamiliar. For our purpose a mapping may be considered simply as a functional relationship. For instance the relation $y = x^2$ maps all numbers x into a parabola (x = ±1, y = +1; x = ±2, y = +4; etc.). The relation y = x maps all numbers x into a straight line making a 45° angle with the y-axis. In these examples one range of numbers is mapped into another range of numbers and we speak of <u>point-functions</u>. The function P(E) is somewhat more complicated; it maps a set onto a range of numbers, and we speak of a set function instead of a point function although the underlying concept is the same. In unsophisticated terms, however, a probability function simply assigns one unique number, a probability, to each event.

Two things should be noted. One is that probability has now been defined without making use of the limit of a ratio. The second thing is that this definition doesn't tell us how to calculate probability; rather, it delineates the mathematical nature of the probability functions. If E is the event "sum = 7," we already know how to calculate its probability assuming that all 36 outcomes in the outcome collection are equally likely:

$$P(E) = \frac{6}{36} = \frac{1}{6}$$

Of course this is a particularly simple example. In other cases we may have to investigate the physical nature of the problem in order to develop the probabilities of events of interest.

9. Bayes' Theorem

The formula of Bayes plays an important and interesting role in the structure of probability theory, and it is of particular significance for us because it illustrates a way of thinking that is characteristic of fault tree analysis. We shall first develop the formula using a set theoretic approach, and then discuss the meaning of the result.

Figure VI-7 portrays the "partitioning" of the universal set Ω into subsets A_1, A_2, A_3, A_4, A_5.

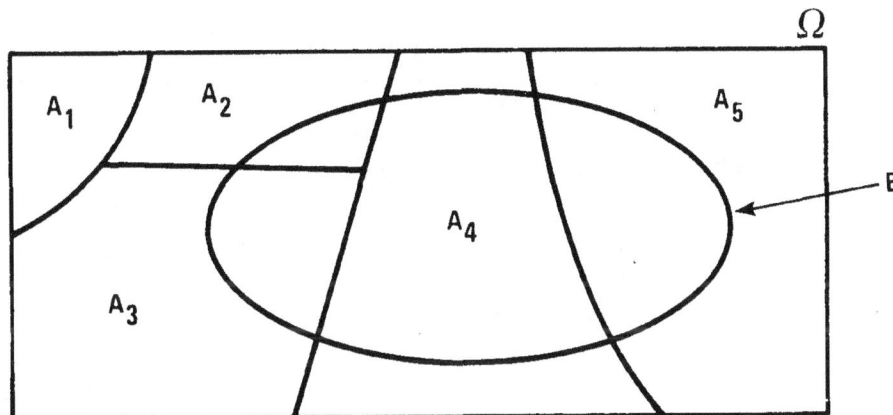

Figure VI-7. Partition of the Universal Set

The A's have the following characteristics:

$$A_1 \cup A_2 \cup A_3 \cup A_4 \cup A_5 = \bigcup_{i=1}^{i=5} A_i = \Omega \qquad \text{(VI-25)}$$

$$A_i \cap A_j = \phi \text{ for } i \neq j.$$

Any set of A's having the properties of Equation (VI-25) is said to constitute a <u>partition</u> of the universal set. Also shown in Figure VI-7 is another subset B. The reader can show (by shading the appropriate regions in the Venn diagram) that

$$(B \cap A_1) \cup (B \cap A_2) \cup (B \cap A_3) \cup (B \cap A_4) \cup (B \cap A_5) = B.$$

(Actually $B \cap A_1 = \phi$ but $\phi \cup X = X$, where X is any set.) The expression for B above can be written in a more mathematically succinct form:

$$B = \bigcup_{i=1}^{i=5} B \cap A_i \qquad \text{(VI-26)}$$

in which the large union symbol implies a succession of unions just as the symbol Σ implies a succession of sums. Similarly, the large intersection symbol implies a succession of intersections just as the symbol Π implies a succession of products. We shall return to (VI-26) in a moment.

Consider now the probability equation for an intersection,

$$P(A \cap B) = P(A|B)\, P(B) = P(B|A)\, P(A).$$

This is true for any arbitrary events A, B. In particular it will be true for B and any one of the A's in Figure VI-7. Thus, we can write

$$P(A_k \cap B) = P(A_k|B)\, P(B) = P(B|A_k)\, P(A_k) \qquad \text{(VI-27)}$$

or

$$P(A_k|B) = \frac{P(A_k \cap B)}{P(B)} = \frac{P(B|A_k)\, P(A_k)}{P(B)} \qquad \text{(VI-28)}$$

We can now write P(B) in a different way by using Equation (VI-26).

$$P(B) = P \left\{ \bigcup_{i=1}^{i=5} B \cap A_i \right\}$$

$$= \sum_{i=1}^{i=5} P(B \cap A_i) = \sum_{i=1}^{i=5} P(B|A_i)\, P(A_i)$$

which can be done because the events $(B \cap A_i)$ are mutually exclusive. If we substitute this expression for P(B) into (VI-28), we obtain

$$P(A_k|B) = \frac{P(B|A_k)\, P(A_k)}{\sum\limits_{i} P(B|A_i)\, P(A_i)}. \qquad \text{(VI-29)}$$

This is Bayes' theorem. The equation is valid in general for any number of events A_i, $A_2, \ldots, A_n$ which are exhaustive and are mutually exclusive (see Equation (VI-25)). The summation then extends from $i = 1$ to $i = n$ instead of $i = 1$ to $i = 5$.

We now discuss some of the meanings of Equation (VI-29). Suppose that some event B has been observed and that we can make a complete list of the mutually exclusive causes of the event B. These causes are just the A's. Notice, however, the restrictions on the A's given by the relations (VI-25).* Now, having observed B, we may be interested in seeking the probability that B was caused by the event A_k. This is what (VI-29) allows us to compute, if we can evaluate all the terms of the right-hand side.

The Bayseian approach is deductive: given a system event, what is the probability of one of its causative factors? This is to be contrasted with the inductive approach: given some particular malfunction, how will the system as a whole behave? The use of Bayes' formula will now be illustrated by a simple example for which it is particularly easy to enumerate the A's.

Suppose that we have three shipping cartons labeled I, II, III. The cartons are all alike in size, shape, and general appearance and they contain various numbers of resistors from companies X, Y, Z as shown in Figure VI-8.

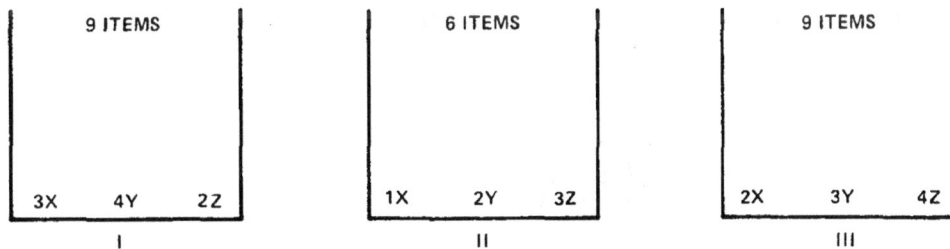

Figure VI-8. Illustration of the Use of Bayes' Formula

A random experiment is conducted as follows: First, one of the cartons is chosen at random. Then two resistors are chosen from the selected box. When they are examined, it is found that both items are from Company Z. This latter event we identify with event B in our general development of Bayes' rule. The "causes" of B are readily identified: either carton I was chosen or carton II was chosen or carton III was chosen. Thus,

A_1 = choice of carton I
A_2 = choice of carton II
A_3 = choice of carton III

Now we should be able to work out the probability that, given event B, it was carton I that was originally chosen.

---

*The restrictions on the A_i given by (VI-25) are equivalent to the restrictions: $\Sigma \ P(A_i) = 1$ and $P(A_i \cap A_j) = 0; i \neq j$.

$$P(A_1 \mid B) = \frac{P(B \mid A_1)\, P(A_1)}{P(B \mid A_1)\, P(A_1) + P(B \mid A_2)\, P(A_2) + P(B \mid A_3)\, P(A_3)}$$

It would appear natural enough to set

$$P(A_1) = P(A_2) = P(A_3) = \frac{1}{3}$$

because the boxes are all alike and a random selection among them is made. We now need to evaluate the terms $P(B \mid A_1)$, $P(B \mid A_2)$ and $P(B \mid A_3)$. This is easily done from Figure VI-8.

$$P(B \mid A_1) = \left(\frac{2}{9}\right)\left(\frac{1}{8}\right) = \frac{1}{36}$$

$$P(B \mid A_2) = \left(\frac{3}{6}\right)\left(\frac{2}{5}\right) = \frac{1}{5}$$

$$P(B \mid A_3) = \left(\frac{4}{9}\right)\left(\frac{3}{8}\right) = \frac{1}{6}.$$

When these numbers are substituted into Bayes' formula we obtain

$$P(A_1 \mid B) = \frac{\left(\frac{1}{36}\right)\left(\frac{1}{3}\right)}{\left(\frac{1}{36}\right)\left(\frac{1}{3}\right) + \left(\frac{1}{5}\right)\left(\frac{1}{3}\right) + \left(\frac{1}{6}\right)\left(\frac{1}{3}\right)} = \frac{5}{71}$$

In a similar way we can calculate

$$P(A_2 \mid B) = \frac{36}{71} \text{ and } P(A_3 \mid B) = \frac{30}{71}.$$

Thus, if event B is actually observed, the chances are about 50-50 that carton II was originally chosen.

As another example, refer once more to our simple system made up of three components. We have already determined that the system can fail in any one of the four modes, $\bar{S}_1, \bar{S}_2, \bar{S}_3, \bar{S}_4$. If the system fails and we wish to know the probability that its failure mode was S_3, we can compute:

$$P(\bar{S}_3 \mid \bar{S}) = \frac{P(\bar{S} \mid \bar{S}_3)\, P(\bar{S}_3)}{P(\bar{S} \mid \bar{S}_1)\, P(\bar{S}_1) + P(\bar{S} \mid \bar{S}_2)\, P(\bar{S}_2) + P(\bar{S} \mid \bar{S}_3)\, P(\bar{S}_3) + P(\bar{S} \mid \bar{S}_4)\, P(\bar{S}_4)}.$$

This can be written in a simpler form because the system will surely fail if any one of the events $\bar{S}_1, \bar{S}_2, \bar{S}_3$, or $\bar{S}_4$ occurs.

$$P(\overline{S}_3 \mid \overline{S}) = \frac{P(\overline{S}_3)}{P(\overline{S}_1) + P(\overline{S}_2) + P(\overline{S}_3) + P(\overline{S}_4)}$$

Presumably the quantities $P(\overline{S}_i)$ can be estimated from reliability data. The quantity $P(\overline{S}_i \mid \overline{S})$ is sometimes called the "importance" of system failure cause $\overline{S}_i$. Bayes' Theorem is sometimes applied to obtain optimal repair schemes and to determine the most likely contributors to the system failure (i.e., which of the $P(\overline{S}_i \mid \overline{S})$ is the largest).

CHAPTER VII – BOOLEAN ALGEBRA AND APPLICATION TO FAULT TREE ANALYSIS

1. Rules of Boolean Algebra

In the previous chapter we developed the elementary theory of sets as applied, specifically, to events (outcomes of random experiments). In this chapter we are going to further develop the algebra of events, called Boolean algebra, with particular application to fault trees. Boolean algebra is especially important in situations involving a dichotomy: switches are either open or closed, valves are either open or closed, events either occur or they do not occur.

The Boolean techniques discussed in this chapter have immediate practical importance in relation to fault trees. A fault tree can be thought of as a pictorial representation of those Boolean relationships among fault events that cause the top event to occur. In fact, a fault tree can always be translated into an entirely equivalent set of Boolean equations. Thus an understanding of the rules of Boolean algebra contributes materially toward the construction and simplification of fault trees. Once a fault tree has been drawn, it can be evaluated to yield its qualitative and quantitative characteristics. These characteristics cannot be obtained from the fault tree per se, but they can be obtained from the equivalent Boolean equations. In this evaluation process we use the algebraic reduction techniques discussed in this chapter.

We present the rules of Boolean algebra in Table VII-1 along with a short discussion of each rule. The reader is urged to check the validity of each rule by recourse to Venn diagrams. Those readers who are mathematically inclined will detect that the rules, as stated, do not constitute a minimal necessary and sufficient set. Here and elsewhere, the authors have sometimes sacrificed mathematical elegance in favor of presenting things in a form that is more useful and understandable for the practical system analyst.

According to (1a) and (1b), the union and intersection operations are commutative. In other words, the commutative laws permit us to interchange the events X, Y with regard to an "AND" operation. It is important to remember that there are mathematical entities that do not commute; e.g. the vector cross product and matrices in general.

Relations (2a) and (2b) are similar to the associative laws of ordinary algebra: $a(bc) = (ab)c$ and $a + (b+c) = (a+b) + c$. If we have a series of "OR" operations or a series of "AND" operations, the associative laws permit us to group the events any way we like.

The distributive laws (3a) and (3b) provide the valid manipulatory procedure whenever we have a combination of an "AND" operation with an "OR" operation. If we go from left to right in the equations, we are simply reducing the left-hand expression to an unfactored form. In (3a), for example, we operate with X on Y and on Z to obtain the right-hand expression. If we go from right to left in the equations, we are simply factoring the expression. For instance, in (3b) we factor out X to obtain the left-hand side. Although (3a) is analogous to the distributive law in ordinary algebra, (3b) has no such analog.

Table VII-2. Rules of Boolean Algebra

Mathematical Symbolism	Engineering Symbolism	Designation
(1a) $X \cap Y = Y \cap X$	$X \cdot Y = Y \cdot X$	Commutative Law
(1b) $X \cup Y = Y \cup X$	$X + Y = Y + X$	
(2a) $X \cap (Y \cap Z) = (X \cap Y) \cap Z$	$X \cdot (Y \cdot Z) = (X \cdot Y) \cdot Z$ $X(YZ) = (XY)Z$	Associative Law
(2b) $X \cup (Y \cup Z) + (X \cup Y) \cup Z$	$X + (Y + Z) = (X + Y) + Z$	
(3a) $X \cap (Y \cup Z) = (X \cap Y) \cup (X \cap Z)$	$X \cdot (Y + Z) = X \cdot Y + X \cdot Z$ $X(Y + Z) = XY + XZ$	Distributive Law
(3b) $X \cup (Y \cap Z) = (X \cup Y) \cap (X \cup Z)$	$X + Y \cdot Z = (X + Y) \cdot (X + Z)$	
(4a) $X \cap X = X$	$X \cdot X = X$	Idempotent Law
(4b) $X \cup X = X$	$X + X = X$	
(5a) $X \cap (X \cup Y) = X$	$X \cdot (X + Y) = X$	Law of Absorption
(5b) $X \cup (X \cap Y) = X$	$X + X \cdot Y = X$	
(6a) $X \cap X' = \phi$	$X \cdot X' = \phi$	Complementation
(6b) $X \cup X' = \Omega = I^*$	$X + X' = \Omega = I$	
(6c) $(X')' = X$	$(X')' = X$	
(7a) $(X \cap Y)' = X' \cup Y'$	$(X \cdot Y)' = X' + Y'$	de Morgan's Theorem
(7b) $(X \cup Y)' = X' \cap Y'$	$(X + Y)' = X' \cdot Y'$	
(8a) $\phi \cap X = \phi$	$\phi \cdot X = \phi$	Operations with ϕ and Ω
(8b) $\phi \cup X = X$	$\phi + X = X$	
(8c) $\Omega \cap X = X$	$\Omega \cdot X = X$	
(8d) $\Omega \cup X = \Omega$	$\Omega + X = \Omega$	
(8e) $\phi' = \Omega$	$\phi' = \Omega$	
(8f) $\Omega' = \phi$	$\Omega' = \phi$	
(9a) $X \cup (X' \cap Y) = X \cup Y$	$X + X' \cdot Y = X + Y$	These relationships are unnamed but are frequently useful in the reduction process.
(9b) $X' \cap (X \cup Y') = X' \cap Y' = (X \cup Y)'$	$X' \cdot (X + Y') = X' \cdot Y' = (X + Y)'$	

*The symbol I is often used instead of Ω to designate the Universal Set. In engineering notation Ω is often replaced by 1 and ϕ by 0.

The idempotent laws (4a) and (4b) allow us to "cancel out" any redundancies of the same event.

The laws of absorption (5a) and (5b) can easily be validated by reference to an appropriate Venn diagram. With respect to (5a), we can also argue in the following way. Whenever the occurrence of X automatically implies the occurrence of Y, then X is said to be a subset of Y. We can symbolize this situation as X⊂Y or X→Y. In this case X+Y = Y and X·Y = X. In (5a), if X occurs then (X+Y) has also occurred and X⊂(X+Y); therefore X·(X+Y) = X. We can develop a similar argument in the case of (5b).

De Morgan's theorems (7a) and (7b) provide the general rules for removing primes on brackets. Suppose that X represents the failure of some component. Then X′

represents the non-failure or successful operation of that component. In this light (7a) simply states that for the double failure of X and Y not to occur, either X must not fail or Y must not fail.

As an application of the use of these rules, let us try to simplify the expression

$(A+B) \cdot (A+C) \cdot (D+B) \cdot (D+C)$.

We can apply (3b) to $(A+B) \cdot (A+C)$ obtaining

$(A+B) \cdot (A+C) = A+(BC)$.

Likewise,

$(D+B) \cdot (D+C) = D+(B \cdot C)$.

We thus have as an intermediate result

$(A+B) \cdot (A+C) \cdot (D+B) \cdot (D+C) = (A+B \cdot C) \cdot (D+B \cdot C)$.

If we now let E represent the event $B \cdot C$, we have

$(A+BC) \cdot (D+BC) = (A+E) \cdot (D+E) = (E+A) \cdot (E+D)$.

Another application of (3b) yields

$(E+A) \cdot (E+D) = E+A \cdot D = B \cdot C + A \cdot D$.

We, therefore, have as our final result

$(A+B) \cdot (A+C) \cdot (D+B) \cdot (D+C) = B \cdot C + A \cdot D$.

The original expression has been substantially simplified for purposes of evaluation.

There are more general rules for simplifying Boolean functions and we will discuss these later in this chapter. For the moment we are concerned with what can be accomplished by more or less unsystematic manipulation of the algebra. A few examples of this procedure are now given. The reader should work carefully through these illustrations and ascertain at each step which of the rules, (1a) - (9b), are being used.

Example 1—Show that

$$[(A \cdot B) + (A \cdot B') + (A' \cdot B')]' = A' \cdot B.$$

This example can be worked either by (a) removing the outermost prime as a first step or by (b) manipulating the terms inside the large brackets and removing the outermost prime as a last step. In either case the removal of primes is accomplished by using (7a) or (7b).

(a) $[(A \cdot B) + (A \cdot B') + (A' \cdot B')]'$
$= (A \cdot B)' \cdot (A \cdot B')' \cdot (A' \cdot B')'$
$= (A'+B') \cdot (A'+B) \cdot (A+B)$
$= A' + (B' \cdot B) \cdot (A+B)$
$= (A'+\phi) \cdot (A+B)$
$= A' \cdot (A+B)$
$= (A' \cdot A) + (A' \cdot B)$
$= \phi + (A' \cdot B)$
$= A' \cdot B.$

(b) $[(A \cdot B) + (A \cdot B') + (A' \cdot B')]'$
$= [(A \cdot (B+B') + (A' \cdot B')]'$
$= [A \cdot \Omega + (A' \cdot B')]'$
$= [A + (A' \cdot B')]'$
$= [(A+A') \cdot (A+B')]'$
$= [\Omega \cdot (A+B')]'$
$= [A+B']' = A' \cdot B.$

Example 2—Show that

$(A' \cdot B \cdot C')' \cdot (A \cdot B' \cdot C')' = C + [(A' \cdot B') + (B \cdot A)]$
$(A' \cdot B \cdot C')' \cdot (A \cdot B' \cdot C')'$
$= (A+B'+C) \cdot (A'+B+C)$
$= C + [(A+B') \cdot (A'+B)]$
$= C + [(A+B') \cdot A' + (A+B') \cdot B]$
$= C + [(A' \cdot A) + (A' \cdot B') + (B \cdot B') + (B \cdot A)]$
$= C + [\phi + (A' \cdot B') + \phi + (B \cdot A)]$
$= C + [(A' \cdot B') + (B \cdot A)] .$

Example 3—Show that

$[(X \cdot Y) + (A \cdot B \cdot C)] \cdot [(X \cdot Y) + (A'+B'+C')] = X \cdot Y$

Applying de Morgan's theorem to the second term inside the second bracket, we have

$[(X \cdot Y) + (A \cdot B \cdot C)] \cdot [(X \cdot Y) + (A \cdot B \cdot C)'].$

Now let L= $X \cdot Y$, M= $(A \cdot B \cdot C)$, and we have

$(L+M) \cdot (L+M') = L \cdot (M+M') = L + \Omega = L = X \cdot Y$

and the original statement is proved. Notice in this example that A,B,C are completely redundant.

2. Application to Fault Tree Analysis

In this section we shall relate the Boolean methodology to fault trees. A fault tree, as we now know, is a logic diagram depicting certain events that must occur in order for other events to occur. The events are termed "faults" if they are initiated by other events and are termed "failures" if they are the basic initiating events. The fault tree interrelates events (faults to faults or faults to failures) and certain symbols are used to depict the various relationships (see Chapter IV). The basic symbol is the "gate" and each gate has inputs and an output as shown in Figure VII-1.

The gate output is the "higher" fault event under consideration and the gate inputs are the more basic ("lower") fault (or failure) events which relate to the output. When we draw a fault tree, we proceed from the "higher" faults to the more basic faults (i.e., from output to inputs). In this process (Chapter V), certain

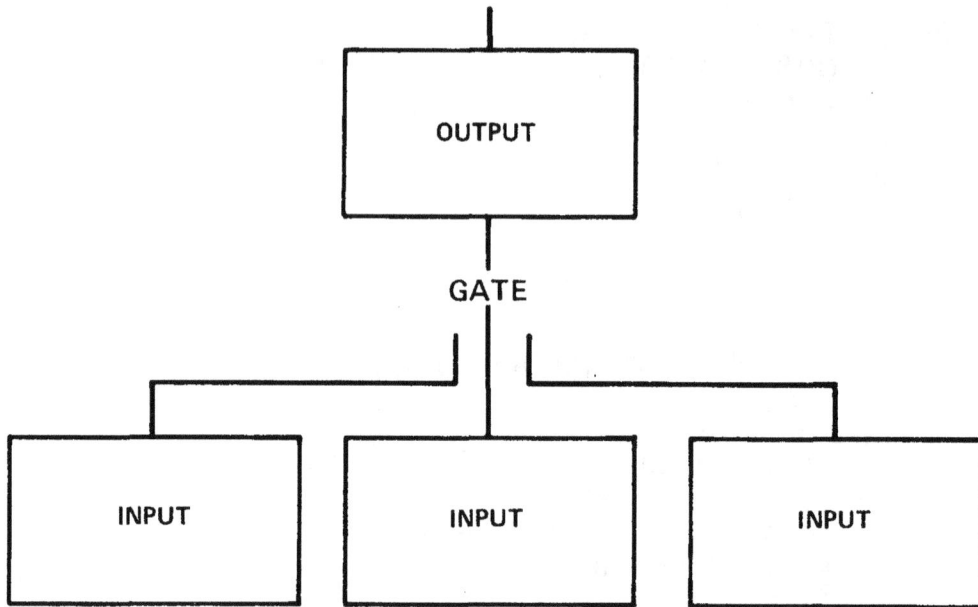

Figure VII-1. The Gate Function in a Fault Tree

techniques are used to determine which category of gate is appropriate. The two basic gate categories are the OR-gate and the AND-gate. Because these gates relate events in exactly the same way as the Boolean operations that we have just discussed, there is a one-to-one correspondence between the Boolean algebraic representation and the fault tree representation.

The OR-Gate

The fault tree symbol ⌂ is an OR-gate which represents the <u>union</u> of the events attached to the gate. Any one or more of the input events must occur to cause the event above the gate to occur. The OR-gate is equivalent to the Boolean symbol "+." For example, the OR-gate with two input events, as shown in Figure VII-2, is equivalent to the Boolean expression, $Q=A+B$. Either of the events A or B, or both must occur in order for Q to occur. Because of its equivalence to the Boolean union operation denoted by the symbol "+," the OR-gate is sometimes drawn with a "+" inside the gate symbol as in Figure VII-2. For n input events attached to the OR-gate, the equivalent Boolean expression is $Q = A_1 + A_2 + A_3 + \ldots + A_n$.
In the terms of probability, from Equation (VI-4):

$$P(Q) = P(A) + P(B) - P(A \cap B)$$
or
$$= P(A) + P(B) - P(A)P(B|A) \qquad \text{(VII-1)}$$

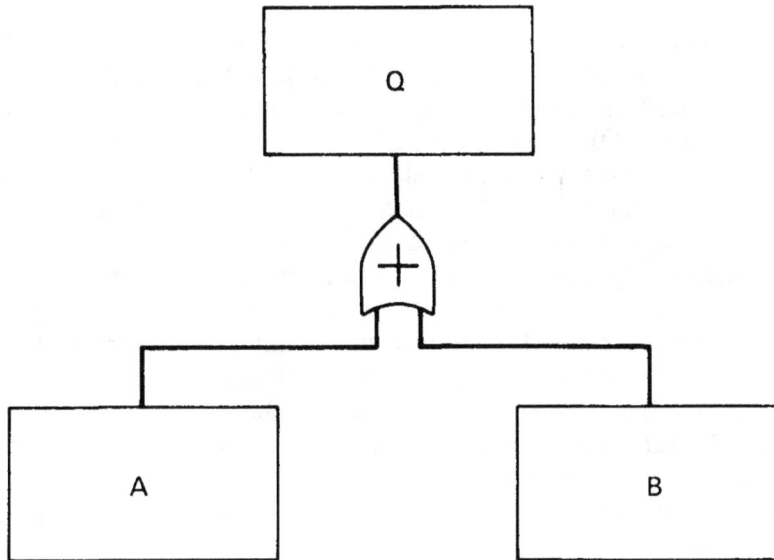

Figure VII-2. A Two-Input OR-Gate

From our discussion of probabilities in Chapter VI, we can make the following observations about Equation (VII-1):

- If A and B are mutually exclusive events, then $P(A \cap B) = 0$ and
  $P(Q) = P(A) + P(B)$;
- If A and B are independent events, then $P(B|A) = P(B)$ and
  $P(Q) = P(A) + P(B) - P(A) P(B)$;
- If event B is completely dependent on event A, that is, whenever A occurs, B also occurs, then $P(B|A) = 1$ and
  $P(Q) = P(A) + P(B) - P(A)$
  $\quad\quad = P(B)$;
- The approximation $P(Q) \cong P(A) + P(B)$ is, in all cases, a conservative estimate for the probability of the output event Q, i.e.,
  $P(A) + P(B) \geqslant P(A) + P(B) - P(A \cap B)$ for all A, B;
- If A and B are independent, low probability events (say $P(A)$, $P(B) < 10^{-1}$), then $P(A \cap B)$ is small compared with $P(A) + P(B)$ so that $P(A) + P(B)$ is a very accurate approximation of $P(Q)$. The reader will remember $P(A) + P(B)$ as the "rare event approximation" discussed in Chapter VI.

These observations, especially the last two, will become very important when we discuss quantification in Chapter XI.

In Chapter IV, we briefly discussed the EXCLUSIVE OR-gate. As the reader should remember, the output event Q of an EXCLUSIVE OR-gate with two input events A and B occurs if event A occurs or event B occurs, but <u>not</u> both. The probability expression for the output event Q of an EXCLUSIVE OR-gate is:

$$P(Q)_{\text{EXCLUSIVE OR}} = P(A) + P(B) - 2P(A \cap B) \quad\quad\quad\quad \text{(VII-2)}$$

Comparing Equations (VII-1) and (VII-2), we observe that if A and B are independent low probability component failures, the difference in probability between the two expressions is negligible. This is why the distinction between the inclusive and exclusive OR-gates is generally not necessary in Fault Tree Analysis where we are often dealing with independent, low probability component failures. It may sometimes, however, be useful to make the distinction in the special case where the exclusive OR logic is truly required, and in addition where there is a strong dependency between the input events and the failure probabilities are high. In this latter case, the intersection term may be large enough to significantly effect the result. In conclusion, it should be observed that in any case, the error which is made by using the inclusive rather than the exclusive OR-gate biases the answer on the conservative side because the inclusive OR has the higher probability. In the remainder of this text, unless otherwise noted, all references to the OR-gate should be interpreted as the inclusive variety.

Figure VII-3 shows a realistic example of an OR-gate for a fault condition of a set of normally closed contacts.

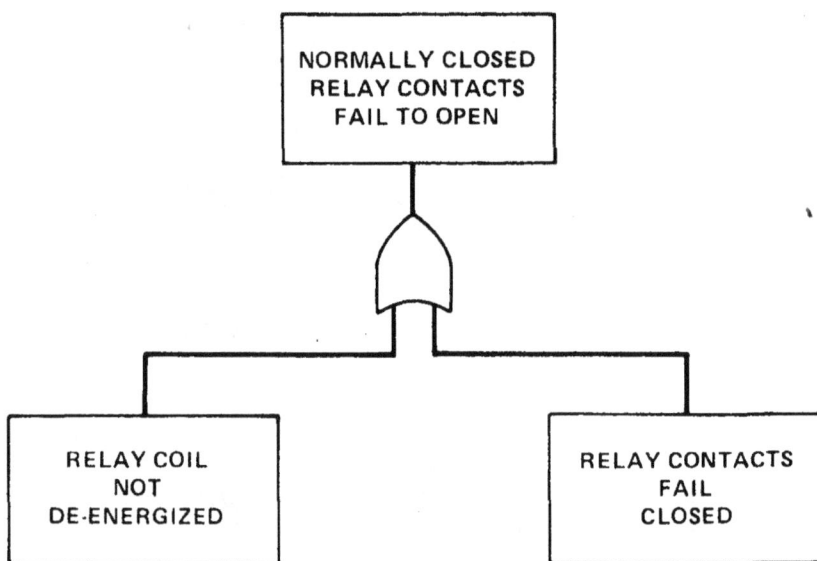

Figure VII-3. A Specific Two-Input OR-Gate

This OR-gate is equivalent to the Boolean expression

Instead of explicitly describing the events, a unique symbol (Q, A_2, etc.) is usually associated with each event as shown in Figure VII-2. Therefore, if the event "relay contacts fail to open" is labeled "Q," "relay coil not de-energized" is labeled "A," and "contacts fail closed" is labeled "B," we can represent the OR-gate of Figure VII-3 by the Boolean equation Q=A+B.

An OR-gate is merely a re-expression of the event above the gate in terms of the more elementary input events. The event above the gate encompasses all of these more elementary events; if any one or more of these elementary events occurs, then B occurs. This interpretation is quite important in that it characterizes an OR-gate and differentiates it from an AND-gate. The input events to an OR-gate do not cause the event above the gate; they are simply re-expressions of the event above the gate. We have discussed this topic before, but we feel it is so important to fault tree analysis that we cover it again, having now gone through the Boolean algebra concepts.

Consider two switches in series as shown in Figure VII-4. The points A and B are points on the wire. If wire failures are ignored then the fault tree representation of the event, "No Current to Point B" is shown in Figure VII-5.

Figure VII-4. Two Switches in Series

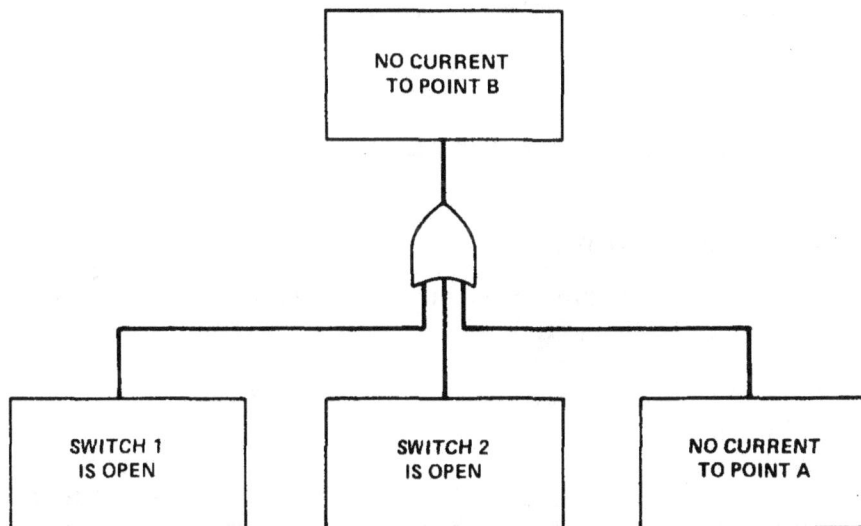

Figure VII-5. A Specific Three-Input OR-Gate

If the events are denoted by the symbols given below, then the Boolean representation is $B = A_1 + A_2 + A_3$.

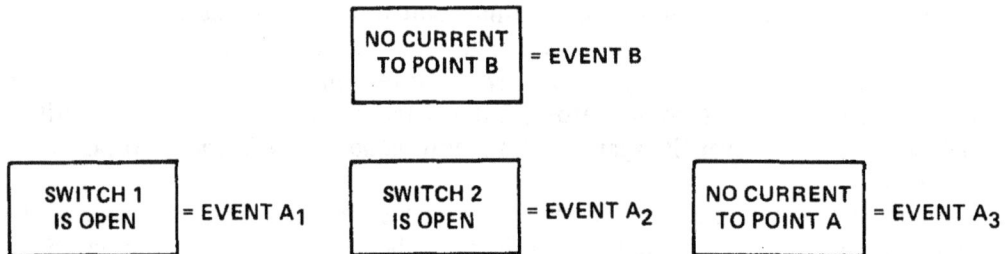

$$\boxed{\begin{array}{c} \text{NO CURRENT} \\ \text{TO POINT B} \end{array}} = \text{EVENT B}$$

$$\boxed{\begin{array}{c} \text{SWITCH 1} \\ \text{IS OPEN} \end{array}} = \text{EVENT } A_1 \qquad \boxed{\begin{array}{c} \text{SWITCH 2} \\ \text{IS OPEN} \end{array}} = \text{EVENT } A_2 \qquad \boxed{\begin{array}{c} \text{NO CURRENT} \\ \text{TO POINT A} \end{array}} = \text{EVENT } A_3$$

The event B occurs if A_1 or A_2 or A_3 occurs. Event B is merely a re-expression of events A_1, A_2, A_3. We have classified the particular events A_1, A_2, A_3 as belonging to the general event B.

The AND-Gate

The fault tree symbol ⌂ is an AND-gate which represents the <u>intersection</u> of the events attached to the gate. The AND-gate is equivalent to the Boolean symbol "·". All of the input events attached to the AND-gate must exist in order for the event above the gate to occur. For two events attached to the AND-gate, the equivalent Boolean expression is $Q = A \cdot B$, as shown in Figure VII-6. Because of its equivalence to the Boolean intersection operation denoted by the symbol "·", that symbol is sometimes included inside the AND-gate as in Figure VII-6. For n input events to an AND-gate, the equivalent Boolean expression is

$$Q = A_1 \cdot A_2 \cdot A_3 \cdot \ldots \cdot A_n.$$

In this case, event Q will occur if and only if all the A_i occur. In terms of probability, from Equation (VI-10):

$$P(Q) = P(A)P(B|A) = P(B)P(A|B). \tag{VII-3}$$

From our discussion of probability in Chapter VI, we can make the following observations about Equation (VII-3):

- If A and B are independent events, then $P(B|A) = P(B)$, $P(A|B) = P(A)$, and $P(Q) = P(A) P(B)$;

- If A and B are not independent events, then $P(Q)$ may be significantly greater than $P(A)P(B)$. For example, in the extreme case where B depends completely on A, that is, whenever A occurs, B also occurs, then $P(B|A) = 1$ and $P(Q) = P(A)$.

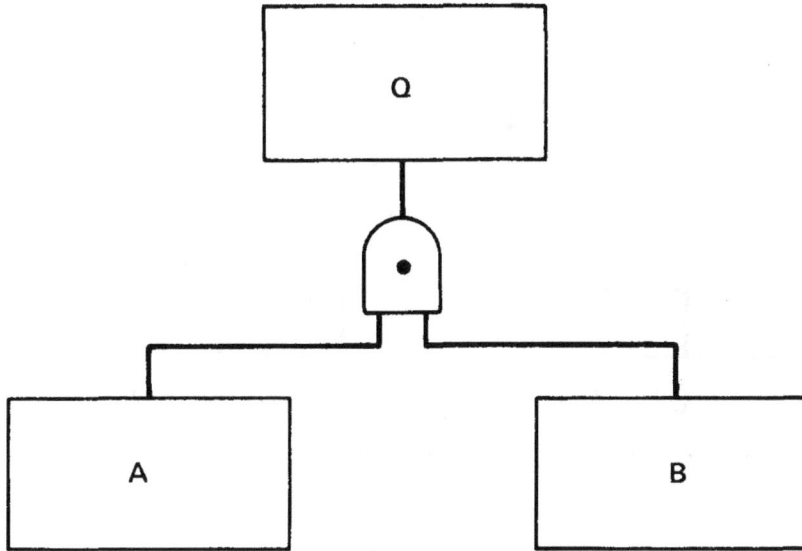

Figure VII-6. A Two-Input AND-Gate

Again, the reader should bear these observations in mind for future reference when we discuss fault tree quantification in later chapters.

The events attached to the AND-gate are the causes of the event above the gate. Event Q is caused only if every one of the input events occurs. This causal relationship is what differentiates an AND-gate from an OR-gate. If the event above the gate occurs when any one of the input events occurs, then the gate is an OR-gate and the event is merely a restatement of the input events. If the event above the gate occurs only when combinations of more elementary events occur, then the gate is an AND-gate and the inputs constitute the cause of the event above the gate.

We conclude this section with an example showing how Boolean algebra can be used to restructure a fault tree. Consider the equation $D = A \cdot (B+C)$. The corresponding fault tree structure is shown in Figure VII-7.

Now according to Rule 3a, event D can also be expressed as $D = (A \cdot B) + (A \cdot C)$. The fault tree structure for this equivalent expression for D is shown in Figure VII-8.

The two fault tree structures in Figures VII-7 and VII-8 may appear to be different; however, they are equivalent. Thus, there is not one "correct" fault tree for a problem but many correct forms which are equivalent to one another. The rules of Boolean algebra can thus be applied to restructure the tree to a simpler, equivalent form for ease of understanding or for simplifying the evaluation of the tree. Later, we shall apply the rules of Boolean algebra to obtain one form of the fault tree, called the minimal cut set form, which allows quantitative and qualitative evaluations to be performed in a straightforward manner.

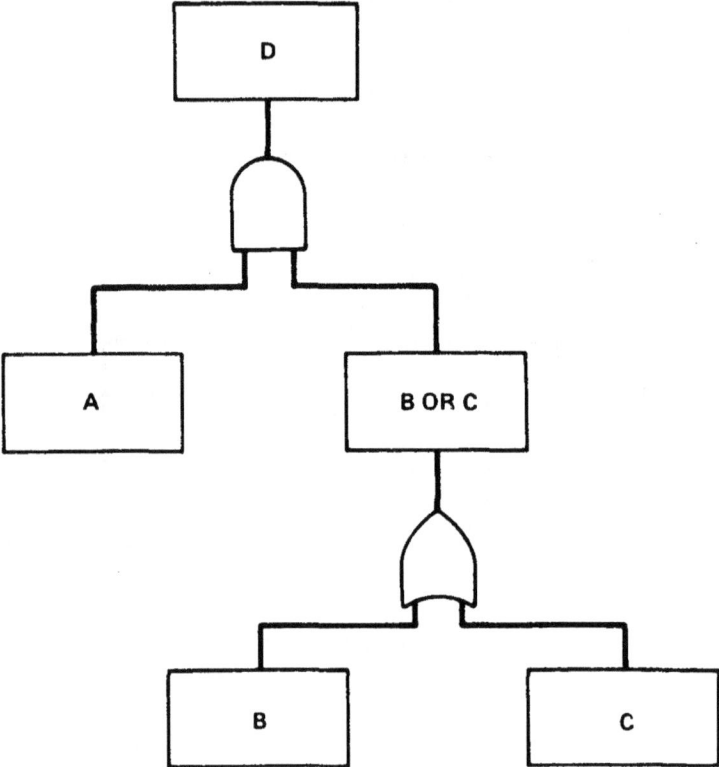

Figure VII-7. Fault Tree Structure for $D = A \cdot (B+C)$

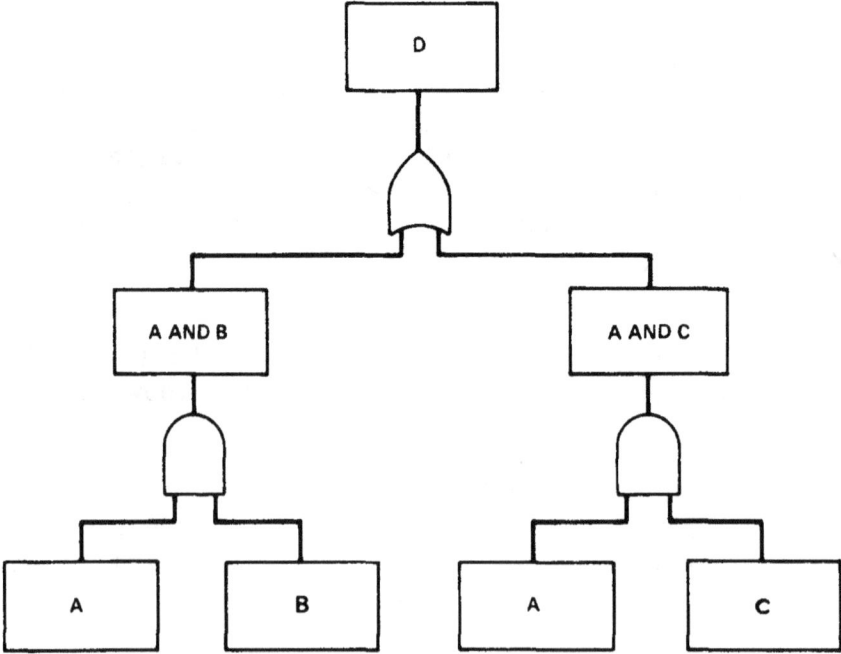

Figure VII-8. Equivalent Form for the Fault Tree of Figure VII-7

3.   Shannon's Method for Expressing Boolean Functions in Standardized Forms

In the previous sections we have discussed how Boolean functions may be expressed in a number of ways by applying the algebraic rules presented earlier in this chapter. In this section we shall discuss Shannon's method for expanding Boolean functions. This method is a general expansion technique applicable to any Boolean function. According to Shannon's method, a Boolean function of n variables can be expanded about one, two, . . ., or all n of the variables. When the expansion is "complete" (i.e., when it is taken about all n of the variables), the result is referred to as an "expansion in minterms." The latter constitutes a standard or canonical form consisting of all combinations of occurrences and non-occurrences of the events of interest.

First some preliminaries need to be reviewed. A Boolean variable is a two-valued variable. For instance if E designates some event of interest, then E=1 indicates that the event has occurred and E=0 indicates that the event has not occurred. For this reason, theorems in Boolean algebra are much more easily proved than theorems in ordinary algebra in which the variable may take on an infinity of values.

Consider a function of the n Boolean variables $X_1, X_2, X_3, \ldots, X_n$:
$f(X_1, X_2, X_3, \ldots, X_n)$.
The Boolean function can take on only two values: 1 (occurs) and 0 (does not occur). This function may be expanded about one of its arguments (say X_1) in the following way:

$$f(X_1, X_2, X_3, \ldots, X_n) = [X_1 \cdot f(1, X_2, \ldots, X_n)] \\ + [X_1' \cdot f(0, X_2, \ldots, X_n)] \qquad \text{(VII-4)}$$

In equation (VII-4), because we are dealing with events, we must remember that the dots (·) and pluses (+) represent the intersection and union operations. The symbolism $f(1, X_2, \ldots, X_n)$ indicates that 1 has been substituted for X_1. We may simplify $f(1, X_2 \ldots, X_n)$ using the rules of Boolean algebra. For example if $f(X_1, X_2) = X_1 \cdot X_2$, then $f(1, X_2) = 1 \cdot X_2 = X_2$. The reader can readily show that equation (VII-4) is correct by considering the only two possibilities, $X_1 = 1$ and $X_1 = 0$, i.e., X_1 occurs or does not occur.

Equation (VII-4) can be extended to the expansion of a Boolean function about two, three, or, indeed, all of its arguments. For instance, the extension of equation (VII-4) to the expansion about the two variables X_1 and X_2 is:

$$f(X_1, X_2, X_3, \ldots, X_n) = [X_1 \cdot X_2 \cdot f(1, 1, X_3, \ldots, X_n)] \\ + X_1 \cdot X_2' \cdot f(1, 0, X_3, \ldots, X_n)] + X_1' \cdot X_2 \cdot f(0, 1, X_3, \ldots, X_n)] \\ + X_1' \cdot X_2' \cdot f(0, 0, X_3, \ldots, X_n)] \qquad \text{(VII-5)}$$

The reader should note that equation (VII-5) can be obtained by simply expanding $f(1, X_2, \ldots, X_n)$ and $f(0, X_2, \ldots, X_n)$ about X_2 in the manner of equation (VII-4). When the expression is carried out about all the variables $X_1, X_2, X_3, \ldots, X_n$, the Boolean expressions lying outside the functional representations are called <u>minterms</u> which consist of combinations of certain X's occurring and others not occurring. In the complete expansion there will be 2^n minterms consisting of all combinations of the possible occurrences and nonoccurrences of the X variables. Each minterm expression will have as a "coefficient" the function f evaluated at the appropriate 1's and 0's corresponding to the occurrences and nonoccurrences of X.

In addition to the minterm expansion, the complementary maxterm expansion can be obtained for any Boolean function. The maxterm expansion is simply obtained from the minterm expansion by interchanging $\cdot$ for $+$ and 0 for 1, and vice versa. Thus from equation (VII-4), the maxterm expansion in one variable is

$$f(X_1, X_2, \ldots, X_n) = [X_1 + f(0, X_1 \ldots, X_n)] \cdot [X'_1 + f(1, X_x, \ldots, X_n)]$$

$$\text{(VII-6)}$$

Analogous forms are obtained for higher order expansions. Instead of obtaining a sum (union) of combinations (intersections) as in the minterm expansion, we obtain a combination (intersection) of sums (unions) for the maxterm expansion.

In the minterm (or maxterm) expansion, each combination term is disjoint* from all the others. Thus in equation (VII-5) the four terms

$$X_1 \cdot X_2 \cdot f(1, 1, X_3, \ldots, X_n)$$
$$X_1 \cdot X'_2 \cdot f(1, 0, X_3, \ldots, X_n)$$
$$X'_1 \cdot X_2 \cdot f(0, 1, X_3, \ldots, X_n)$$
$$X'_1 \cdot X'_2 \cdot f(0, 0, X_3, \ldots, X_n)$$

are disjoint from one another. This characteristic is generally true for any order expansion and it can be exploited in quantifying fault trees whenever Shannon's expansion is used. Another reason for representing a Boolean function in either minterms or maxterms is that these expressions are unique for any given function. Such an expansion, then, provides a general technique for determining whether two Boolean expressions are equal, because if they are, they will have identical minterm (or maxterm) forms.

Shannon's expansion will now be illustrated for the two 3-variable functions:

(a) $f(X,Y,Z) = (X \cdot Y) + (X' \cdot Z) + (Y \cdot Z)$
(b) $f(X,Y,Z) = (X \cdot Y) + (X' \cdot Z)$.

The reader can show, in a few simple algebraic steps, that the functions in (a) and (b) are indeed equal but we wish to use (a) and (b) to exemplify the general process involved in using Shannon's expansion. We shall start with (a) and expand it using the minterm expansion.

$$(X \cdot Y) + (X' \cdot Z) + (Y \cdot Z)$$
$$= [X \cdot Y \cdot Z \cdot f(1, 1, 1)] + [X \cdot Y \cdot Z' \cdot f(1, 1, 0)] + [X \cdot Y' \cdot Z \cdot f(1, 0, 1)]$$
$$+ [X \cdot Y' \cdot Z' \cdot f(1, 0, 0)] + [X' \cdot Y \cdot Z \cdot f(0, 1, 1)] + [X' \cdot Y \cdot Z' \cdot f(0, 1, 0)]$$
$$+ [X' \cdot Y' \cdot Z \cdot f(0, 0, 1)] + [X' \cdot Y' \cdot Z' \cdot f(0, 0, 0)].$$

$$\text{(VII-7)}$$

---

*Two events are disjoint if their intersection is the null set, or equivalently, two events are disjoint if the probability of both events is zero.

Note that this expression is valid for any 3-variable Boolean function. Now f(1, 1, 1), f(1, 1, 0), etc., can be readily evaluated by making the appropriate substitutions in the original functional form as follows:

$$f(1,1,1) = (1\cdot1) + (0\cdot1) + (1\cdot1) = 1+0+1 = 1$$
$$f(1,1,0) = (1\cdot1) + (0\cdot0) + (1\cdot0) = 1+0+0 = 1$$
$$f(1,0,1) = (1\cdot0) + (0\cdot1) + (0\cdot1) = 0+0+0 = 0$$
$$f(1,0,0) = (1\cdot0) + (0\cdot0) + (0\cdot0) = 0+0+0 = 0$$
$$f(0,1,1) = (0\cdot1) + (1\cdot1) + (1\cdot1) = 0+1+1 = 1$$
$$f(0,1,0) = (0\cdot1) + (1\cdot0) + (1\cdot0) = 0+0+0 = 0$$
$$f(0,0,1) = (0\cdot0) + (1\cdot1) + (0\cdot1) = 0+1+0 = 1$$
$$f(0,0,0) = (0\cdot0) + (1\cdot0) + (0\cdot0) = 0+0+0 = 0$$

When these values are substituted into the expanded form of (a) above, the result is the unique minterm expansion of (a):

$$(X\cdot Y) + (X'\cdot Z) + (Y\cdot Z) =$$
$$(X\cdot Y\cdot Z) + (X\cdot Y\cdot Z') + (X'\cdot Y\cdot Z) + (X'\cdot Y'\cdot Z).$$

in which all of the terms in parentheses are disjoint.

It now remains to evaluate f(1,1,1), f(1,1,0), etc., for expression (b).

$$f(1,1,1) = (1\cdot1) + (0\cdot1) = 1+0 = 1$$
$$f(1,1,0) = (1\cdot1) + (0\cdot0) = 1+0 = 1$$
$$f(1,0,1) = (1\cdot0) + (0\cdot1) = 0+0 = 0$$
$$f(1,0,0) = (1\cdot0) + (0\cdot0) = 0+0 = 0$$
$$f(0,1,1) = (0\cdot1) + (1\cdot1) = 0+1 = 1$$
$$f(0,1,0) = (0\cdot1) + (1\cdot0) = 0+0 = 0$$
$$f(0,0,1) = (0\cdot0) + (1\cdot1) = 0+1 = 1$$
$$f(0,0,0) = (0\cdot0) + (1\cdot0) = 0+0 = 0$$

It is now apparent that (b) has exactly the same minterm expansion as (a) and therefore we have proved that $(X\cdot Y) + (X'\cdot Z) + (Y\cdot Z) = (X\cdot Y) + (X'\cdot Z)$.

It is of interest to view the minterms of equation (VII-7) with the help of Venn diagrams. This is shown in Figure VII-9. The reader should note from Figure VII-9 two important properties of the minterms: They are all mutually disjoint and their collective union yields the universal set.

The maxterm expansion of a 3-variable Boolean function can be obtained from an extension of equation (VII-6) as follows:

$$f(X, Y, Z)$$
$$= [X+Y+Z+f(0,0,0)] \cdot [X+Y+Z'+f(0,0,1)] \cdot [X+Y'+Z+f(0,1,0)]$$
$$\cdot [X+Y'+Z'+f(0,1,1)] \cdot [X'+Y+Z+f(1,0,0)] \cdot [X'+Y+Z'+f(1,0,1)]$$
$$\cdot [X'+Y'+Z+f(1,1,0)] \cdot [X'+Y'+Z'+f(1,1,1)]$$

$$(VII-8)$$

The reader should now be able to show that the maxterm expansions of (a) and (b) are identical.

$$X \cdot Y \cdot Z \qquad X \cdot Y \cdot Z' \qquad X \cdot Y' \cdot Z \qquad X \cdot Y' \cdot Z'$$

$$X' \cdot Y \cdot Z \qquad X' \cdot Y \cdot Z' \qquad X' \cdot Y' \cdot Z \qquad X' \cdot Y' \cdot Z'$$

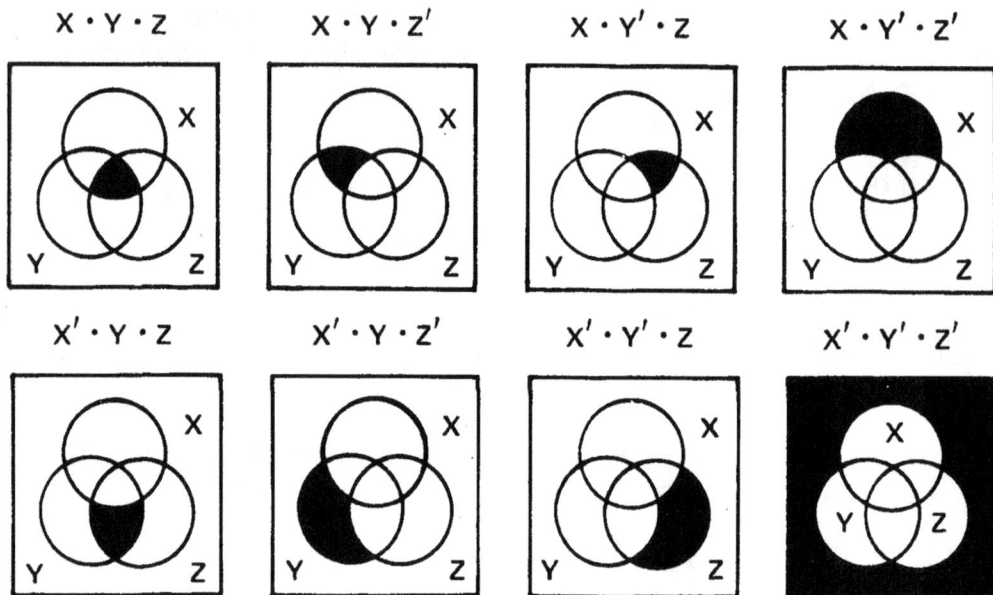

Figure VII-9. Venn Diagram Representation of the Minterms of a
3-Variable Boolean Function

4. Determining the Minimal Cut Sets or Minimal Path Sets of a Fault Tree

One of the main purposes of representing a fault tree in terms of Boolean equations is that these equations can then be used to determine the fault tree's associated "minimal cut sets" and "minimal path sets." The minimal cut sets define the "failure modes" of the top event and are usually obtained when a fault tree is evaluated. Once the minimal cut sets are obtained, the quantification of the fault tree is more or less straightforward. The minimal path sets are essentially the complements of the minimal cut sets and define the "success modes" by which the top event will not occur. The minimal path sets are often not obtained in a fault tree evaluation; however, they can be useful in particular problems.

Minimal Cut Sets

We can formally define a <u>minimal cut set</u> as follows: a minimal cut set is a smallest combination of component failures which, if they all occur, will cause the top event to occur.

By the definition, a minimal cut set is thus a combination (intersection) of primary events sufficient for the top event. The combination is a "smallest" combination in that all the failures are needed for the top event to occur; if one of the failures in the cut set does not occur, then the top event will not occur (by this combination).

Any fault tree will consist of a finite number of minimal cut sets, which are unique for that top event. The one-component minimal cut sets, if there are any, represent those single failures which will cause the top event to occur. The two-component minimal cut sets represent the double failures which together will

cause the top event to occur. For an n-component minimal cut set, all n components
in the cut set must fail in order for the top event to occur.

The minimal cut set expression for the top event can be written in the general
form,

$$T = M_1 + M_2 + \ldots + M_k$$

where T is the top event and M_1 are the minimal cut sets. Each minimal cut set
consists of a combination of specific component failures, and hence the general
n-component minimal cut can be expressed as

$$M_2 = X_1 \cdot X_2 \ldots X_n$$

where X_1, X_2, etc., are basic component failures on the tree. An example of a top
event expression is

$$T = A + B \cdot C$$

where A, B, and C are component failures. This top event has a one-component minimal cut set (A) and a two-component minimal cut set (B·C). The minimal cut sets
are unique for a top event and are independent of the different equivalent forms the
same fault tree may have.

To determine the minimal cut sets of a fault tree, the tree is first translated to its
equivalent Boolean equations and then either the "top-down" or "bottom-up"
substitution method is used. The methods are straightforward and they involve
substituting and expanding Boolean expressions. Two Boolean laws, the distributive
law and the law of absorption, are used to remove the redundancies.

Consider the simple fault tree shown in Figure VII-10; the equivalent Boolean
equations are shown below the tree.

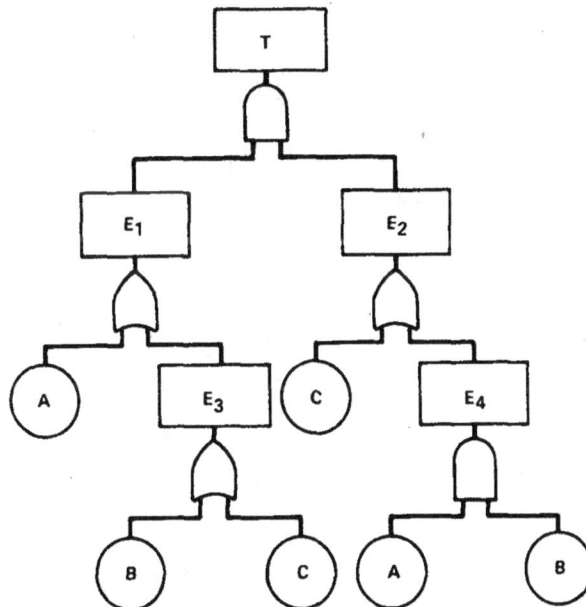

Figure VII-10. Example Fault Tree

$$T = E_1 \cdot E_2$$
$$E_1 = A + E_3$$
$$E_3 = B + C$$
$$E_2 = C + E_4$$
$$E_4 = A \cdot B$$

We will first perform the top-down substitution. We start with the top event equation and substitute and expand until the minimal cut set expression for the top event is obtained. Substituting for E_1 and E_2 and expanding we have:

$$T = (A + E_3) \cdot (C + E_4)$$
$$= (A \cdot C) + (E_3 \cdot C) + (E_4 \cdot A) + (E_3 \cdot E_4)$$

Substituting for E_3:

$$T = A \cdot C + (B + C) \cdot C + E_4 \cdot A + (B + C) \cdot E_4$$
$$= A \cdot C + B \cdot C + C \cdot C + E_4 \cdot A + E_4 \cdot B + E_4 \cdot C.$$

By the idempotent law, $C \cdot C = C$, so we have:

$$T = A \cdot C + B \cdot C + C + E_4 \cdot A + E_4 \cdot B + E_4 \cdot C.$$

But $A \cdot C + B \cdot C + C + E_4 \cdot C = C$ by the law of absorption. Therefore,

$$T = C + E_4 \cdot A + E_4 \cdot B.$$

Finally, substituting for E_4 and applying the law of absorption twice

$$T = C + (A \cdot B) \cdot A + (A \cdot B) \cdot B$$
$$= C + A \cdot B.$$

The minimal cut sets of the top event are thus C and A·B, one single component minimal cut set and one double component minimal cut set. The fault tree can thus be represented as shown in Figure VII-11 which is equivalent to the original tree (both trees have the same minimal cut sets).

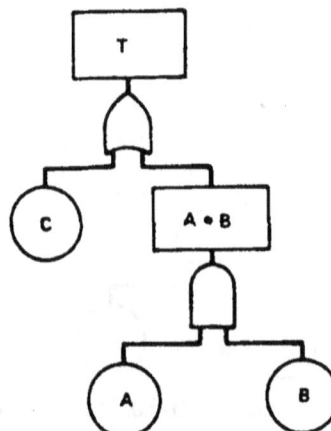

Figure VII-11. Fault Tree Equivalent of Figure VII-10

The bottom-up method uses the same substitution and expansion techniques, except that now the operation begins at the bottom of the tree and proceeds upward. Equations containing only basic failures are successively substituted for higher faults. The bottom-up approach can be more laborious and time-consuming; however, the minimal cut sets are now obtained for every intermediate fault as well as the top event.

Consider again our example tree (we repeat the equivalent Boolean equations for the reader's convenience).

$$T \;= E_1 \cdot E_2$$
$$E_1 = A + E_3$$
$$E_3 = B + C$$
$$E_2 = C + E_4$$
$$E_4 = A \cdot B$$

Because E_4 has only basic failures, we substitute into E_2 to obtain

$$E_2 = C + A \cdot B.$$

The minimal cut sets of E_2 are thus C and $A \cdot B$. E_3 is already in reduced form having minimal cut sets B and C. Substituting into E_1, we obtain $E_1 = A + B + C$ so E_1 has three minimal cut sets A, B, and C. Finally, substituting the expressions for E_1 and E_2 into the equation for T, expanding and applying the absorption law, we have

$$T = (A+B+C) \cdot (C+A \cdot B)$$
$$\;\; = A \cdot C + A \cdot A \cdot B + B \cdot C + B \cdot A \cdot B + C \cdot C + C \cdot A \cdot B$$
$$\;\; = A \cdot C + A \cdot B + B \cdot C + A \cdot B + C + A \cdot B \cdot C$$
$$\;\; = C + A \cdot B.$$

The minimal cut sets of the top event are thus again C and $A \cdot B$.

As a very simple example, suppose we have the pumping system shown in Figure VII-12.

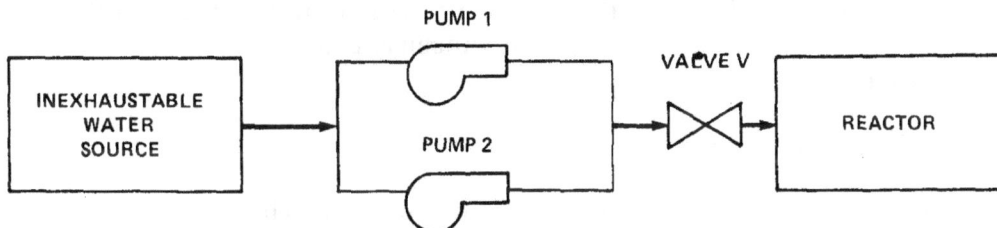

Figure VII-12. Water Pumping System

Assume that out undesired event is "no flow of water to reactor." Ignoring the contribution of the pipes, we can model this system by fault tree of Figure VII-11 where:

T = "no flow of water to reactor"
C = "valve V fails closed"
A = "pump 1 fails to run"
B = "pump 2 fails to run"

We have just shown that the minimal cut sets of this tree are C and A·B. This tells us that our undesired event "no flow of water to tank" will occur if either valve V fails closed or both pumps fail to run. In this simple case, the cut sets do not really provide any insights that are not already quite obvious from the system diagram. In more complex systems, however, where the system failure modes are not so obvious, the minimal cut set computation provides the analyst with a thorough and systematic method for identifying the basic combinations of component failures which can cause an undesired event.

For smaller fault trees, the determination of the minimal cut sets, using either the top-down or bottom-up method, can be done by hand. For larger trees, various computer algorithms and codes for fault tree evaluation are available. These are discussed in Chapter XII.

Minimal Path Sets and Dual Fault Trees

The top event of a fault tree represents system failure. This event is of great interest from the point of view of system safety. From the point of view of reliability we would be more concerned with the prevention of the top event. Now we know that the top event of a fault tree may be represented by a Boolean equation, and because this equation may be complemented, there is also a Boolean equation for the complement (i.e., nonoccurrence) of the top event. This complemented equation, in turn, corresponds to a tree which is the complement of the original tree. This complemented tree, called the dual of the original fault tree, may be obtained directly from the original tree by complementing all the events and substituting OR-gates for AND-gates and vice versa. In either case, whether we complement the top event equation or the tree itself, we are applying de Morgan's theorem given in our Table of Boolean Algebra Laws. The minimal cut sets of the dual tree are the so-called "minimal path sets" of the original tree, where a minimal path set is a smallest combination (intersection) of primary events whose non-occurrence assures the non-occurrence of the top event.

The combination is a smallest combination in that all the primary event nonoccurrences are needed for the top event to not occur; if any one of the events occurs then the top event can occur. The minimal path set expression for the top event T can be written as

$$T' = P_1 + P_2 + \ldots + P_k$$

where T' denotes the complement (nonoccurrence) of T. The terms $P_1, P_2, \ldots, P_x$ are the minimal path sets of the fault tree. Each path set can be written as

$$P_i = X'_1 \cdot X'_2 \cdot \ldots \cdot X'_m$$

where the X_i are the basic events in the fault tree and the X'_i are the complements.

We can find the minimal path sets of a given tree by forming its dual and then using either the top-down or bottom-up method to find its minimal cut sets. These cut sets are the minimal path sets of the original tree which we desire.

Alternatively, if the minimal cut sets of the tree have already been determined, we can take the complement of the minimal cut set equation and obtain the minimal path sets directly. For our sample tree, we obtained the following minimal cut expression in the previous section,

$$T = C + A \cdot B.$$

Taking the complement

$$T' = (C + A \cdot B)'$$
$$= C' \cdot (A \cdot B)'$$

using de Morgan's theorem. Applying de Morgan's theorem to the term $(A \cdot B)'$

$$T' = C' \cdot (A' + B')$$

and using the distributive law (i.e., expanding)

$$T' = C' \cdot A' + C' \cdot B'$$

Therefore, the minimal path sets of the tree are $C' \cdot A'$ and $C' \cdot B'$. In terms of our pumping system of Figure VII-12, this tells us that we can <u>prevent</u> the undesired event and assure system success if either

(1) Valve V is open and pump 1 is running, or
(2) Valve V is open and pump 2 is running.

CHAPTER VIII — THE PRESSURE TANK EXAMPLE

1. System Definition and Fault Tree Construction

In this and the next chapter, we are going to define undesired events for two simple systems. The reader will then be shown how the corresponding fault trees are developed step-by-step with the help of the rules described in Chapter V. From the trees so constructed, some obvious conclusions will be drawn, but detailed evaluation procedures will be postponed until Chapter XI.

Consider now Figure VIII-1 which shows a pressure tank - pump-motor device and its associated control system. First we present details of system operation. The operational modes are given in Figure VIII-2.

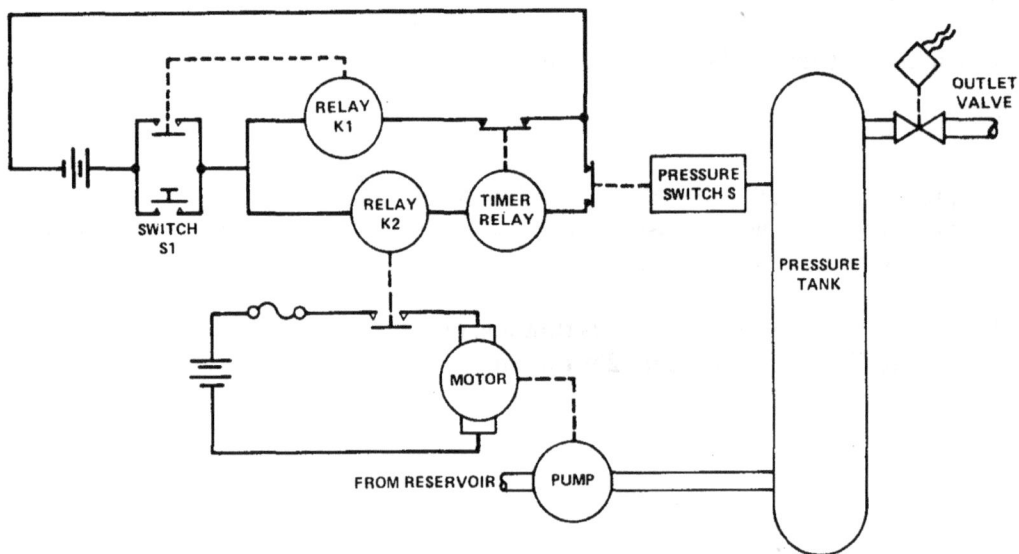

Figure VIII-1. Pressure Tank System

The function of the control system is to regulate the operation of the pump. The latter pumps fluid from an infinitely large reservoir into the tank. We shall assume that it takes 60 seconds to pressurize the tank. The pressure switch has contacts which are closed when the tank is empty. When the threshold pressure has been reached, the pressure switch contacts open, deenergizing the coil of relay K2 so that relay K2 contacts open, removing power from the pump, causing the pump motor to cease operation. The tank is fitted with an outlet valve that drains the entire tank in an essentially negligible time; the outlet valve, however, is not a pressure relief valve. When the tank is empty, the pressure switch contacts close, and the cycle is repeated.

Initially the system is considered to be in its dormant mode: switch S1 contacts open, relay K1 contacts open, and relay K2 contacts open; i.e., the control system is

TRANSITION TO PUMPING

RELAY #2   – ENERGIZED
                  (CLOSED)
TIMER REL   – STARTS TIMING
PRESSURE SW – CONTACTS CLOSE
              PUMP STARTS

READY MODE

S1   – CONTACTS OPEN
RELAY #1   – CONTACTS CLOSED
RELAY #2   – CONTACTS OPEN
TIMER REL   – CONTACTS CLOSED
PRESSURE SW – CONTACTS OPEN
              & MONITORING

TRANSITION TO READY

RELAY #2   – DE-ENERGIZED
                  (OPEN)
TIMER REL   – RESETS TO ZERO
                  TIME
PRESSURE SW – CONTACTS OPEN
              PUMP STOPS

PUMPING MODE

S1   – CONTACTS OPEN
RELAY #1   – CONTACTS CLOSED
RELAY #2   – CONTACTS CLOSED
TIMER REL   – CONTACTS CLOSED
                  & TIMING
PRESSURE SW – CONTACTS CLOSED
              & MONITORING

RELAY #1   – CONTACTS OPEN
RELAY #2   – CONTACTS OPEN
TIMER REL   – TIMES OUT &
                  MOMENTARILY
                  OPENS
PRESSURE SW – FAILED CLOSED
              PUMP STOPS

START-UP TRANSITION

S1   – MOMENTARILY
                  CLOSED
RELAY #1   – ENERGIZED &
                  LATCHED
RELAY #2   – ENERGIZED
                  (CLOSED)
TIMER REL   – STARTS TIMING
PRESSURE SW – MONITORING
                  PRESSURE
              PUMP STARTS

EMERGENCY
SHUTDOWN
TRANSITION
(ASSUME PRESSURE
SWITCHING-UP)

EMERGENCY SHUTDOWN

S1   – CONTACTS OPEN
RELAY #1   – CONTACTS OPEN
RELAY #2   – CONTACTS OPEN
TIMER REL   – CONTACTS CLOSED
PRESSURE SW – CONTACTS CLOSED

DORMANT MODE

SN S1   – CONTACTS OPEN
RELAY #1   – CONTACTS OPEN
RELAY #2   – CONTACTS OPEN
TIMER REL   – CONTACTS CLOSED
PRESSURE SW – CONTACTS CLOSED

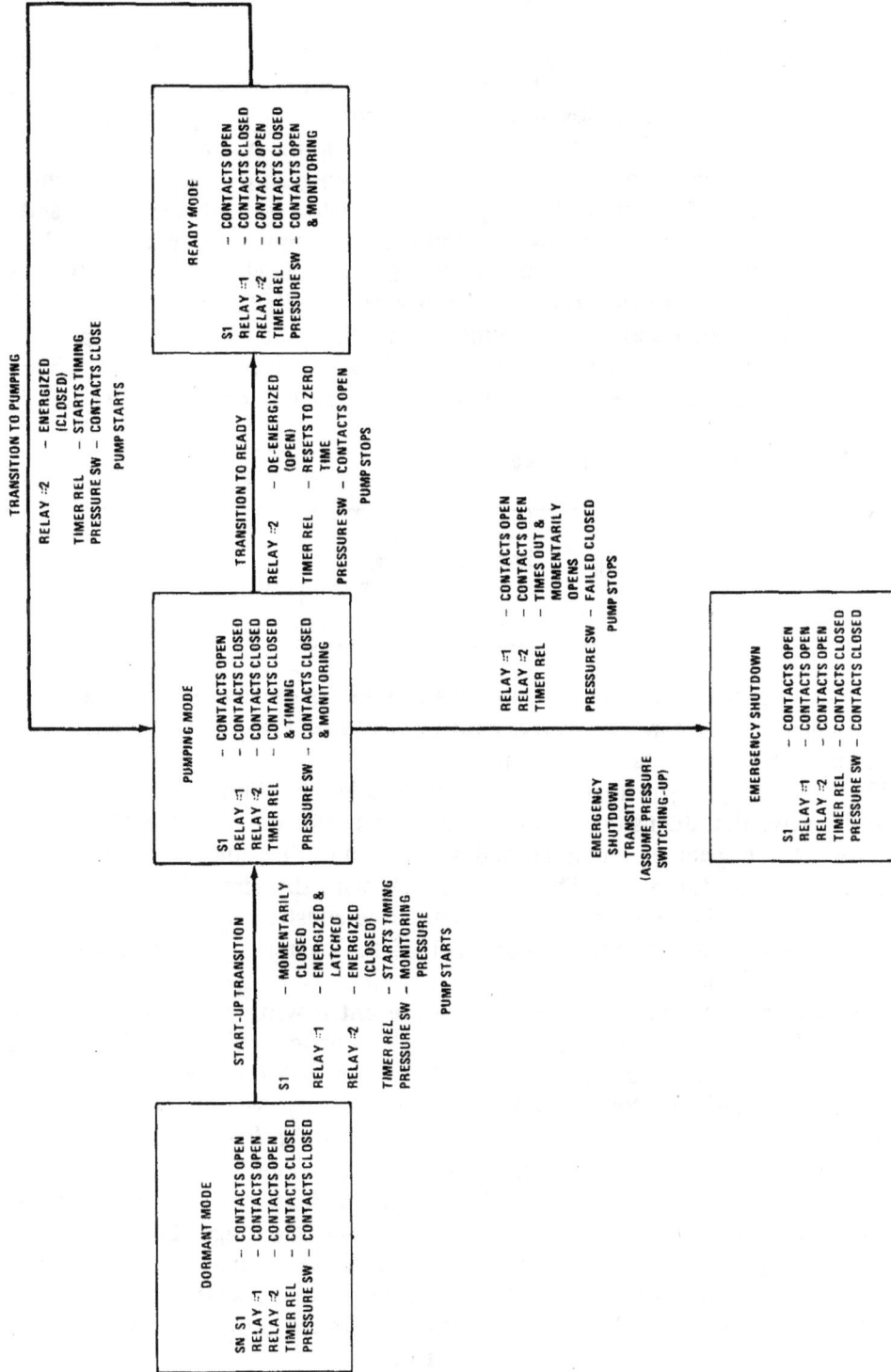

Figure VIII-2. Pressure Tank Example

de-energized. In this de-energized state the contacts of the timer relay are closed. We will also assume that the tank is empty and the pressure switch contacts are therefore closed.

System operation is started by momentarily depressing switch S1. This applies power to the coil of relay K1, thus closing relay K1 contacts. Relay K1 is now electrically self-latched. The closure of relay K1 contacts allows power to be applied to the coil of relay K2, whose contacts close to start up the pump motor.

The timer relay has been provided to allow emergency shut-down in the event that the pressure switch fails closed. Initially the timer relay contacts are closed and the timer relay coil is de-energized. Power is applied to the timer coil as soon as relay K1 contacts are closed. This starts a clock in the timer. If the clock registers 60 seconds of <u>continuous</u> power application to the timer relay coil, the timer relay contacts open (and latch in that position), breaking the circuit to the K1 relay coil (previously latched closed) and thus producing system shut-down. In normal operation, when the pressure switch contacts open (and consequently relay K2 contacts open), the timer resets to 0 seconds.

For our undesired event let us take:

```
┌─────────────────────────┐
│       RUPTURE OF        │
│     PRESSURE TANK       │
│     AFTER THE START     │
│       OF PUMPING        │
└─────────────────────────┘
```

It will simplify things considerably if we agree to neglect plumbing and wiring failures and also all secondary failures except, of course, the one of principal interest: "tank rupture after the start of pumping."

The reader may object that a system that includes an infinitely large reservoir and an outlet valve that drains the tank in a negligible time is unrealistic—and now we are suggesting the neglect of plumbing and wiring faults that might contribute to the occurrence of the top event. The point is that with this simplified system we can illustrate most of the important steps in fault tree construction. In a more complex system the reader might tend to lose sight of the overall system and become too involved in the details.

First we check to make sure that our top event is written as a fault and that it specifies a "what" and a "when." Next we apply our test question: "Can this fault consist of a component failure?" Because the answer is "Yes," we immediately add an OR-gate beneath the top event and consider primary, secondary, and command modes.* Our tree has now developed to the point shown in Figure VIII-3.

In this problem we shall establish our limit of resolution at the "component failure level." By "component" we shall mean those items specifically named in Figure VIII-1. Thus the primary failure of the tank (e.g., a fatigue failure of the tank wall) is already at the limit of resolution and is shown in a circle. Whether or not we include the statement in the diamond is moot. We could just assume at the beginning that the tank was an appropriate one for the operating pressures involved. At any rate, we choose not to trace this fault any further.

---

*In this case, however, there is no command mode.

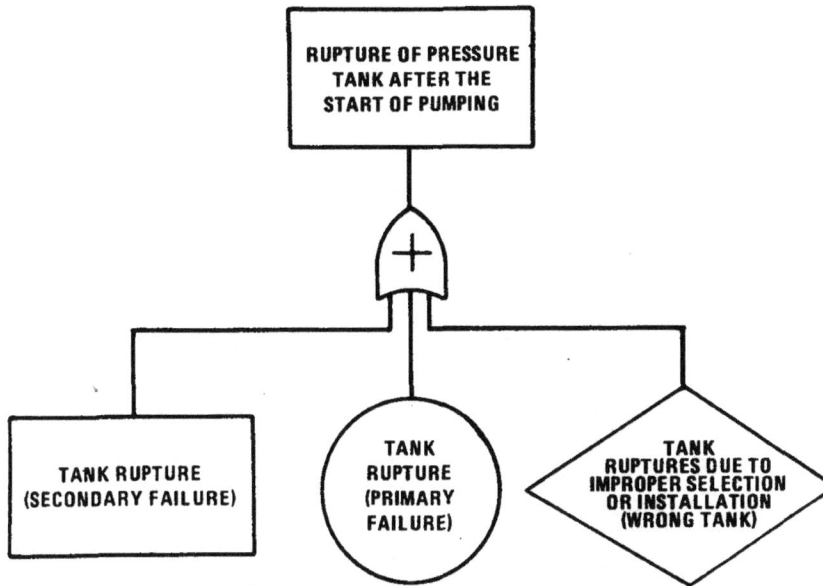

Figure VIII-3. Fault Tree Construction — Step 1

Thus our attention is now directed to the secondary failure of the tank. The reader will remember from Chapter V, that in contrast to a primary failure, which is the failure of a component in an environment for which it is qualified, a secondary failure is the failure of a component in an environment for which it is <u>not</u> qualified. Because the secondary failure of the tank can consist of a component failure, we introduce another OR-gate and our tree assumes the form shown in Figure VIII-4.

Here again we indicate, in a diamond, a set of conditions whose causes we choose not to seek. Notice that the fault spelled out in the rectangle is a specific case of the top event with a more detailed description as to cause.

Now it might happen that our tank could miraculously withstand continuous pumping for t > 60 seconds but an application of our "No Miracles" rule constrains us to the statement that the tank will always rupture under these conditions. We can indicate this on the fault tree by using an Inhibit gate whose input is "continuous pump operation for t > 60 seconds" (see Figure VIII-5).

Can the input event to the Inhibit gate consist of a component failure? No, the pump is simply operating and pump operation for any length of time cannot consist of a component failure. Therefore this fault event must be classified "state-of-system." We now recall the rules of Chapter V. Below a state-of-system fault we can have an OR-gate, an AND-gate, or no gate at all. Furthermore, we look for the minimum, immediate, necessary and sufficient cause or causes. In this case the immediate cause is "motor runs for t > 60 seconds," a state-of-system fault. Its immediate cause is "power applied to motor for t > 60 seconds," a state-of-system fault. The immediate cause of the latter event is "K2 relay contacts remain closed for t > 60 seconds." We have now added the string of events shown in Figure VIII-6.

In this case, nothing is lost by jumping from "pump operates continuously for t > 60 seconds" directly to "K2 relay contacts remain closed for t > 60 seconds." There is, however, no harm done in detailing the intermediate causes and, as a matter of fact, the opportunity for error is lessened thereby.

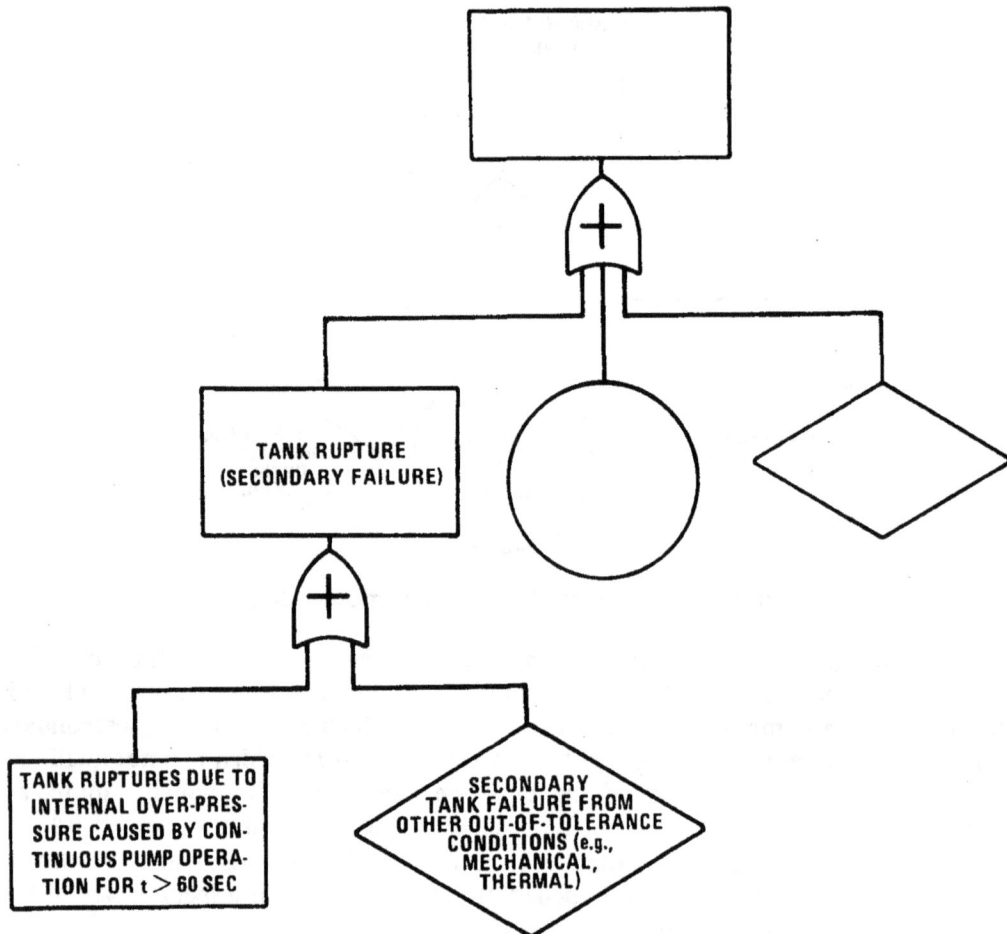

Figure VIII-4. Fault Tree Construction – Step 2

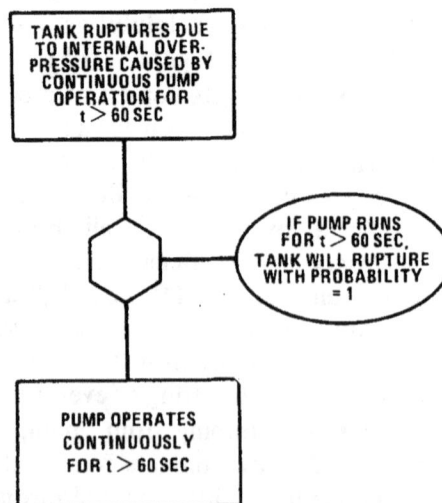

Figure VIII-5. Fault Tree Construction – Step 3

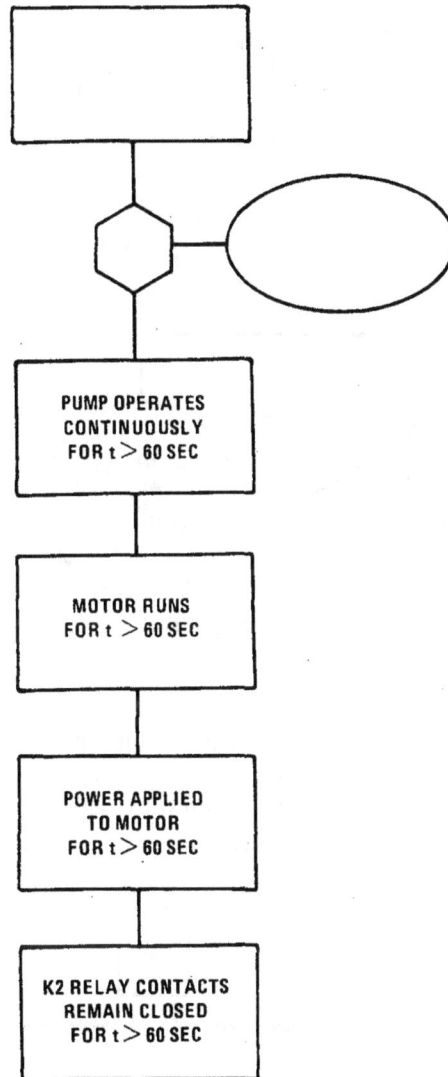

Figure VIII-6. Fault Tree Construction — Step 4

We have now to consider the fault event, "K2 relay contacts closed for t > 60 seconds." Can this consist of a component failure? Yes, the contacts could jam, weld, or corrode shut. We thus draw an OR-gate and add primary, secondary, and command modes as shown in Figure VIII-7.

The event of interest here is the command mode event described in the rectangle. Recall that a command fault involves the proper operation of a component, but in the wrong place or at the wrong time because of an erroneous command of signal from another component. In this case, the erroneous signal is the application of EMF to the relay coil for more than 60 seconds. This state-of-system fault can be analyzed as shown in Figure VIII-8.

Notice that both input events to the AND-gate in Figure VIII-8 are written as faults. In fact, as we know, all events that are linked together on a fault tree should

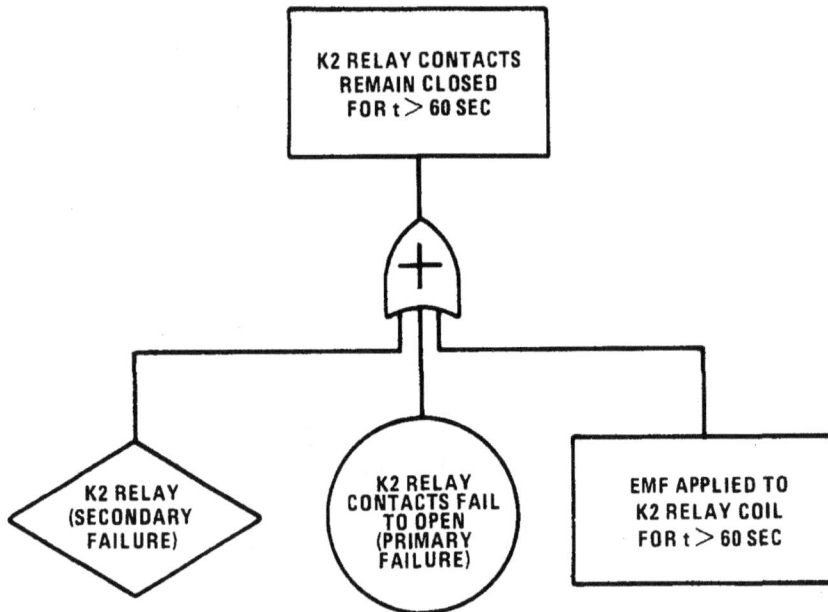

Figure VIII-7. Fault Tree Construction – Step 5

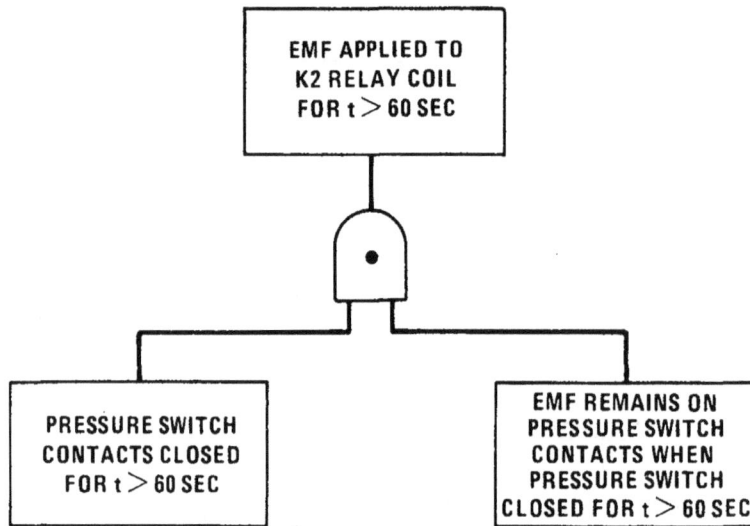

Figure VIII-8. Fault Tree Construction – Step 6

be written as faults except, perhaps, those statements that are added as simply remarks (e.g., statements in ellipses). The pressure switch contacts being closed is not a fault per se, but when they are closed for greater than 60 seconds, that is a fault. Likewise the fact that an EMF is applied to the pressure switch contacts is not itself a fault. Notice that the condition that makes this event a fault is framed in terms of the other input event to the AND-gate.

The fault event, "pressure switch contacts closed for t > 60 seconds," can consist of a component failure, so both input events in Figure VIII-8 are followed by OR-gates. We analyze them separately, starting with the left-hand event (see Figure VIII-9.

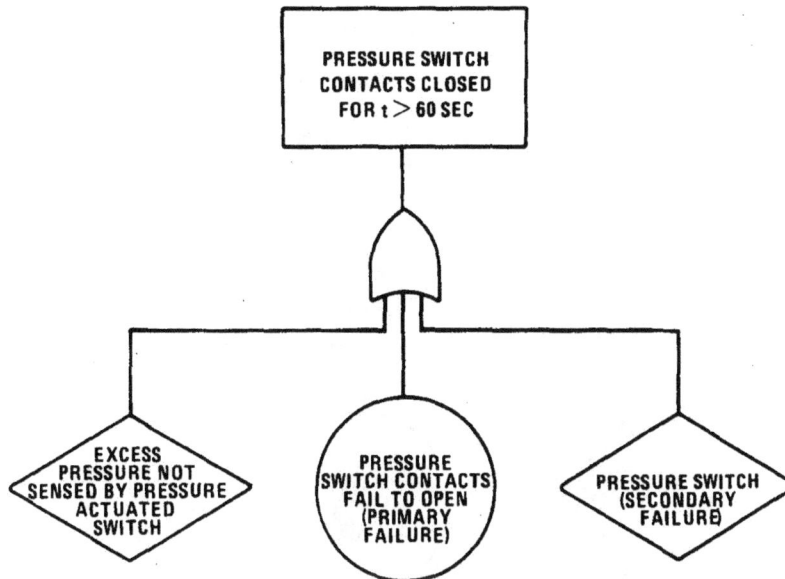

Figure VIII-9. Fault Tree Construction — Step 7

We see that this leg of the tree has reached its terminus (all input events are either circles or diamonds) unless, for some reason, we wish to pursue the event in the left-hand diamond somewhat further (e.g., ruptured diaphragm, etc.).

We now analyze the right-hand event in Figure VIII-8 as shown in Figure VIII-10.

Both the input events in Figure VIII-10 are state-of-component faults. The left-hand one is the more easily analyzed as shown in Figure VIII-11.

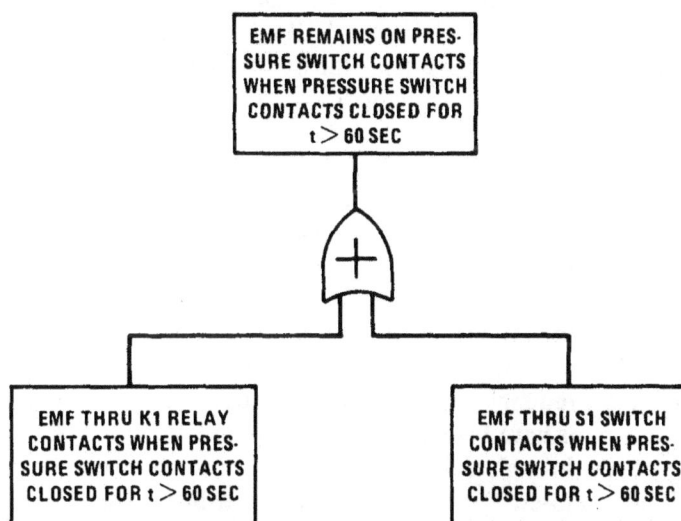

Figure VIII-10. Fault Tree Construction — Step 8

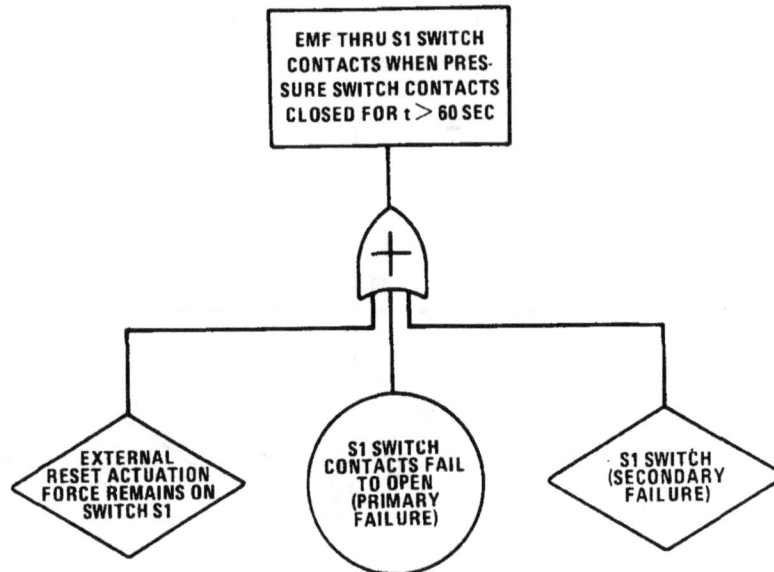

Figure VIII-11. Fault Tree Construction — Step 9

Here we have reached another tree terminus. The analysis of the remaining input event in Figure VIII-10 is shown in Figure VIII-12. The reader will note from this latter figure that we have finally worked our way—in this step-by-step fashion—down to timer relay faults. Finally, we show the complete fault tree for the pressure tank example in Figure VIII-13.

Actually, the fault tree of Figure VIII-13 could be considered _too_ complete. Because the only secondary failure that we developed was the rupture of the pressure tank due to overpumping, other secondary failures (the dotted diamonds) could simply be omitted from the diagram. Further simplifications can also be made, leading to the basic fault tree of Figure VIII-14, where the circles represent primary failures as shown in the legend and the fault events E1, E2 etc. are defined as follows:

The E's are fault events.

- E1 — Pressure tank rupture (top event).
- E2 — Pressure tank rupture due to internal overpressure from pump operation for t > 60 seconds which is equivalent to K2 relay contacts closed for t > 60 seconds.
- E3 — EMF on K2 relay coil for t > 60 seconds.
- E4 — EMF remains on pressure switch contacts when pressure switch contacts have been closed for t > 60 seconds.
- E5 — EMF through K1 relay contacts when pressure switch contacts have been closed for t > 60 seconds which is equivalent to timer relay contacts failing to open when pressure switch contacts have been closed for t > 60 seconds.

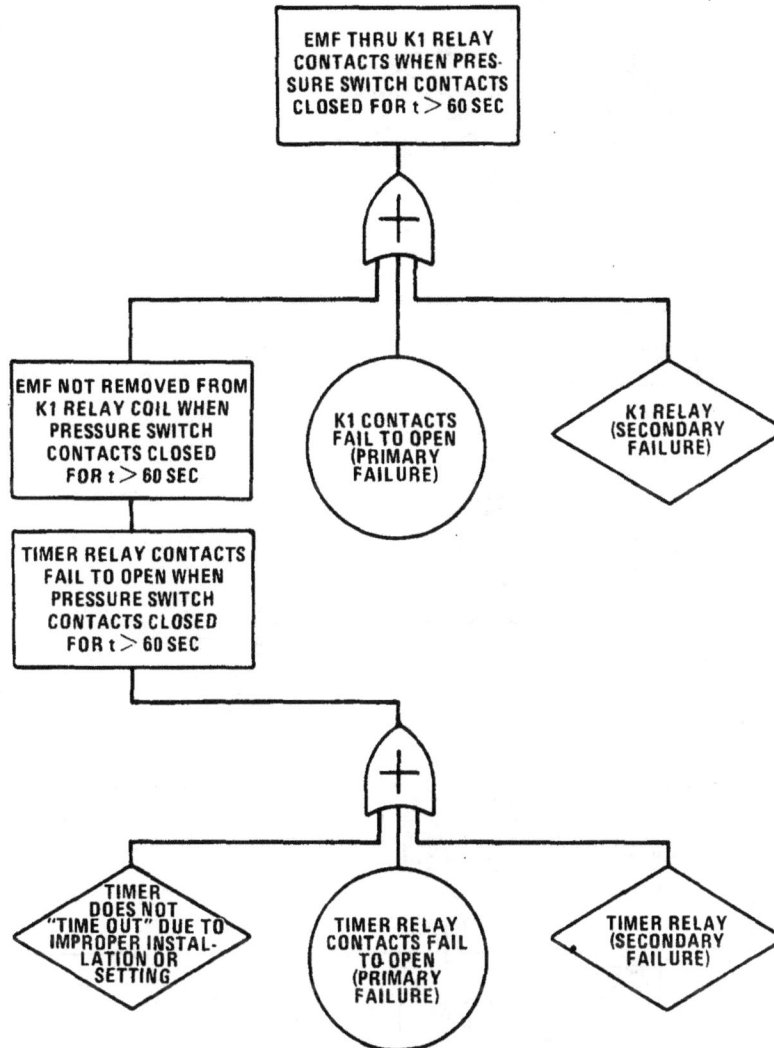

Figure VIII-12. Fault Tree Construction — Final Step

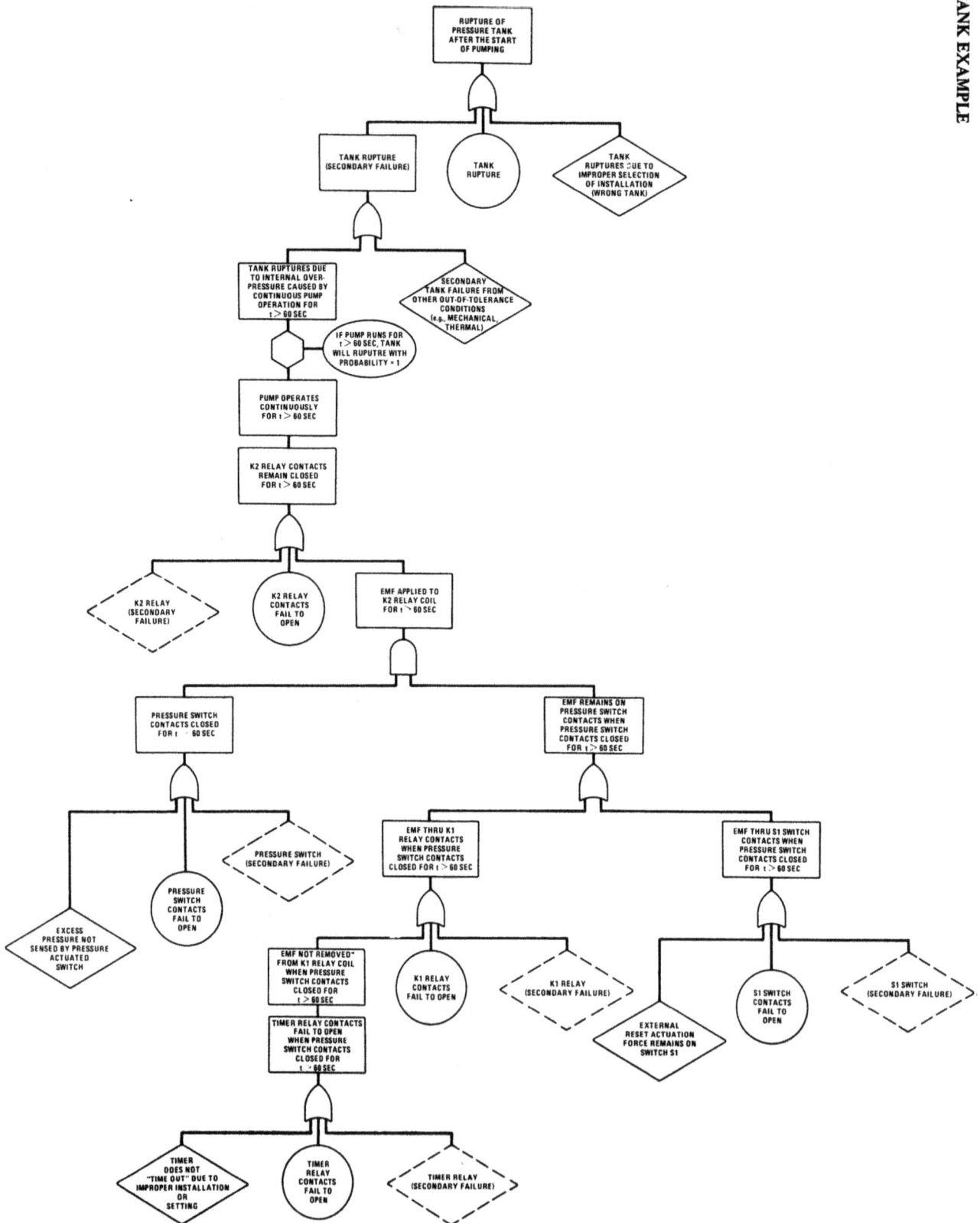

Figure VIII-13. Pressure Tank Rupture Fault Tree Example

2.   Fault Tree Evaluation (Minimal Cut Sets)

Fault tree evaluation in general will be taken up in Chapter XI. It is useful at this point, however, to evaluate the fault tree so that we can assess, in a gross way, the strengths and particularly the weaknesses of the pressure tank control circuit.

From the basic fault tree of Figure VIII-14 we can express the top event as a Boolean function of the primary input events using the method explained in Chapter VII. This is accomplished by starting at the top of the tree and working down:

$$
\begin{aligned}
E1 &= T + E2 \\
&= T + (K2 + E3) \\
&= T + K2 + (S \cdot E4) \\
&= T + K2 + S \cdot (S1 + E5) \\
&= T + K2 + (S \cdot S1) + (S \cdot E5) \\
&= T + K2 + (S \cdot S1) + S \cdot (K1 + R) \\
&= T + K2 + (S \cdot S1) + (S \cdot K1) + (S \cdot R)
\end{aligned}
$$

This expression of the top event in terms of the basic inputs to the tree is the Boolean algebraic equivalent of the tree itself. E1 appears as the union of various combinations (intersections) of basic events and is the minimal cut expression for the top event. In our example, we have found five minimal cut sets—two singles and three doubles:

K2
T
$S \cdot S1$
$S \cdot K1$
$S \cdot R$

Each of these defines an event or series of events whose existence or joint existence will initiate the top event of the tree.

We are now in a position to make, first, a qualitative assessment of our results and then, armed with some data, a gross quantitative assessment. Qualitatively, the leading contributor to the top event is the single relay K2 because it represents a primary failure of an active component. Therefore, the safety of our system would be considerably enhanced by substituting a pair of relays in parallel for the single relay K2. Actually, however, our system contains a much more serious design error: We are monitoring the controls instead of the parameter of interest (pressure in this case). It should be just the other way around! Thus, the most obvious way to improve the system would be to install a pressure relief valve on the tank and remove the timer.

The next basic event, in order of importance, is T, the primary failure of the pressure tank itself. Because the tank is a passive component, recall from Chapter V that the probability of the event T should be less (by an order of magnitude or so) than the probability of event K2. Of less importance are the three double component cut sets $S \cdot S1$, $S \cdot K1$, and $S \cdot K2$, although we do note that the failure of the pressure switch contributes to each of them.

To make a quantitative assessment of our results we need estimates of failure probabilities for our components. Table VIII-1 gives values for the failure probabilities for the components in the system. The calculations involved in getting these numbers will be discussed later; at this time we simply assume the values as being "data."

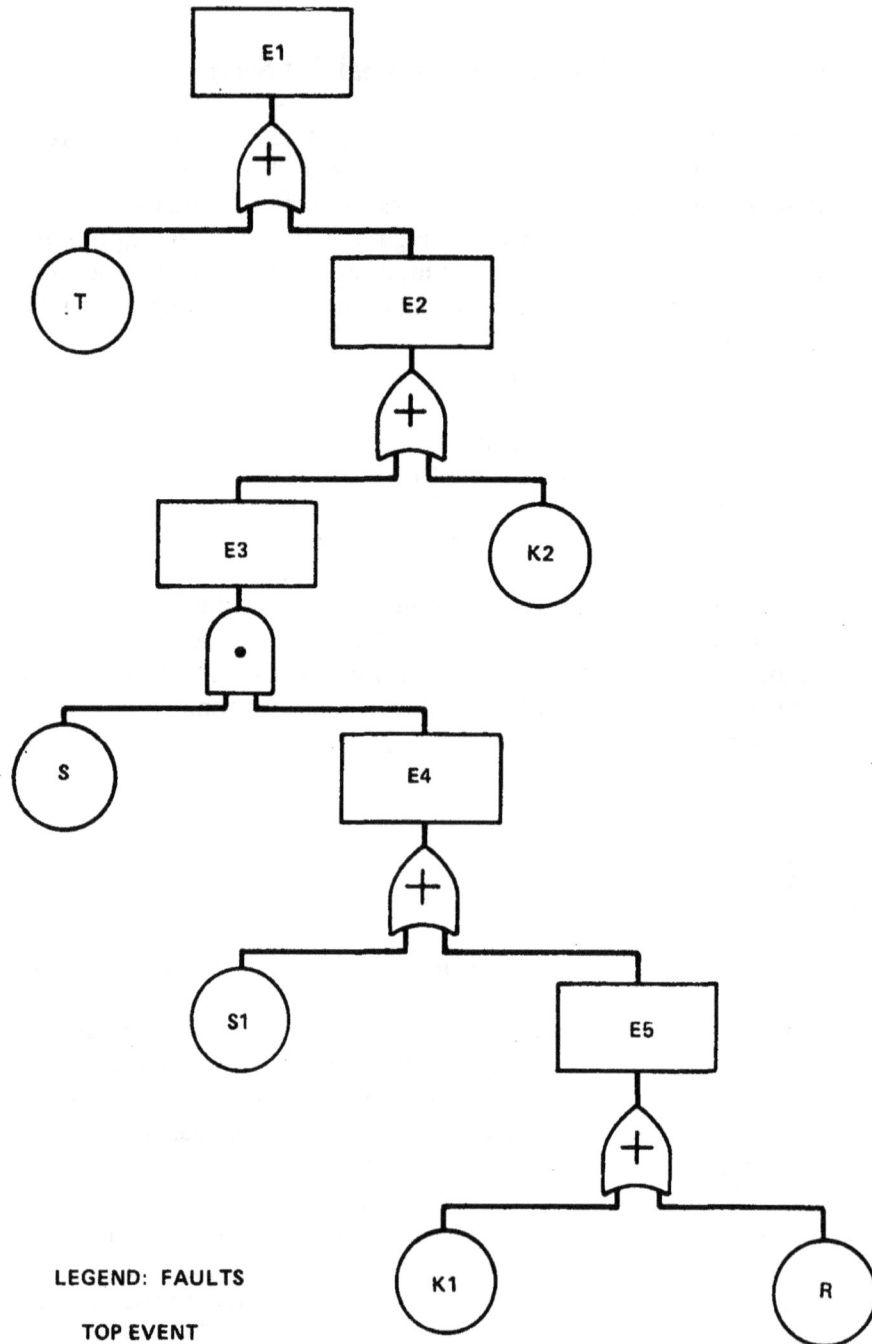

LEGEND: FAULTS

E1	TOP EVENT
E2, E3, E4, E5	INTERMEDIATE FAULT EVENTS
R	PRIMARY FAILURE OF TIMER RELAY
S	PRIMARY FAILURE OF PRESSURE SWITCH
S1	PRIMARY FAILURE OF SWITCH S1
K1	PRIMARY FAILURE OF RELAY K1
K2	PRIMARY FAILURE OF RELAY K2
T	PRIMARY FAILURE OF PRESSURE TANK

Figure VIII-14. Basic (Reduced) Fault Tree
for Pressure Tank Example

Table VIII-1. Failure Probabilities for Pressure Tank Example

COMPONENT	SYMBOL	FAILURE PROBABILITY
Pressure Tank	T	5×10^{-6}
Relay K2	K2	3×10^{-5}
Pressure Switch	S	1×10^{-4}
Relay K1	K1	3×10^{-5}
Timer Relay	R	1×10^{-4}
Switch S1	S1	3×10^{-5}

Because a minimal cut is an intersection of events, the probabilities associated with our five minimal cut sets are obtained by multiplying the appropriate component failure probabilities (assuming independence of failures);

$$
\begin{aligned}
P[T] &= 5 \times 10^{-6} \\
P[K2] &= 3 \times 10^{-5} \\
P[S \cdot K1] &= (1 \times 10^{-4})(3 \times 10^{-5}) = 3 \times 10^{-9} \\
P[S \cdot R] &= (1 \times 10^{-4})(1 \times 10^{-4}) = 1 \times 10^{-8} \\
P[S \cdot S1] &= (1 \times 10^{-4})(3 \times 10^{-5}) = 3 \times 10^{-9}
\end{aligned}
$$

We now wish to estimate the probability of the top event, E_1. Observing that the top event probability is given by the probability of the union of the minimal cut sets, and that the probability of each individual minimal cut set is low, we conclude that the rare event approximation of equation (VI-7) is applicable. We therefore simply sum the minimal cut set probabilities and obtain:

$P(E_1) \cong 3.4 \times 10^{-5}$

The relative quantitative importance of the various cut sets can be obtained by taking the ratio of the minimal cut set probability to the total system probability:

Cut Set	Importance
T	14%
K2	86%
S · K1	
S · R	Less than 0.1%
S · S1	

The pressure tank example has been provided on numerous occasions as a workshop exercise for students learning the basics of fault tree construction. The most frequent analytical error made by these students is the tendency to leap directly from the event of rupture to the pressure switch. When this is done the single failure minimal cut set K2, i.e., the primary failure of K2, is missed completely. This illustrates the importance of applying the rules described in Chapter V.

One design-related principle that arises out of this example is that a system should be designed so that AND-gates appear as close to the tree-top as possible in an effort to eliminate single event cut sets. The family automobile is a good example of a system which is not of this type, and the reader can amuse himself by making a lengthy list of single events that will immobilize his car.

CHAPTER IX – THE THREE MOTOR EXAMPLE

1. System Definition and Fault Tree Construction

In this chapter we discuss a somewhat more complicated example of fault tree construction and evaluation.* Figure IX-1 displays a power distribution box and Figure IX-2 gives the system modus operandi. With contacts KT1, KT2, and KT3 normally closed, a momentary depression of pushbutton S1 applies power from Battery 1 to the coils of cut-throat relays K1 and K2. Thereupon K1 and K2 close and remain electrically latched.

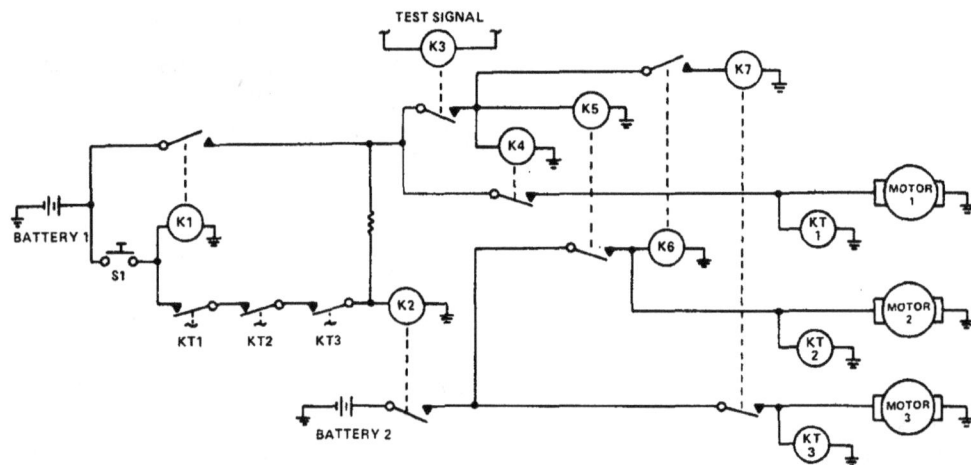

Figure IX-1. Power Distribution Box Fault Tree Example

Next, a 60-second test signal is impressed through K3, the purpose being to check the proper operation of Motors 1, 2, and 3. Once K3 has closed, power from Battery 1 is applied to the coils of relays K4 and K5. The closure of K4 starts Motor 1. The closure of K5 applies power from Battery 2 to the coil of K6 and also starts Motor 2. Finally, the closure of K6 applies power from Battery 1 to the coil of K7. Closure of K7 starts Motor 3.

After an interval of 60 seconds, K3 is supposed to open, shutting down the operation of all three motors. Should K3 fail closed after the expiration of 60 seconds, all three timers (KT1, KT2, KT3) open, de-energizing the coil of K1, thus shutting down system operation. Suppose K3 opens properly at the end of 60 seconds, but K4 fails closed. In that case KT1 opens to deenergize K1 and Motor 1 stops. KT2 and KT3 act similarly to stop Motor 2 or Motor 3 should either K5 or K7 fail closed.

---

*For an actual example of a fault tree analysis applied to a major nuclear safety system see U.S. Nuclear Regulatory Commission, "Reactor Safety Study–An Assessment of Accident Risks in U.S. Commercial Nuclear Power Plants," WASH-1400 (NUREG-75/014), Appendix II, "Fault Trees," Section 2, "Fault Tree Analysis."

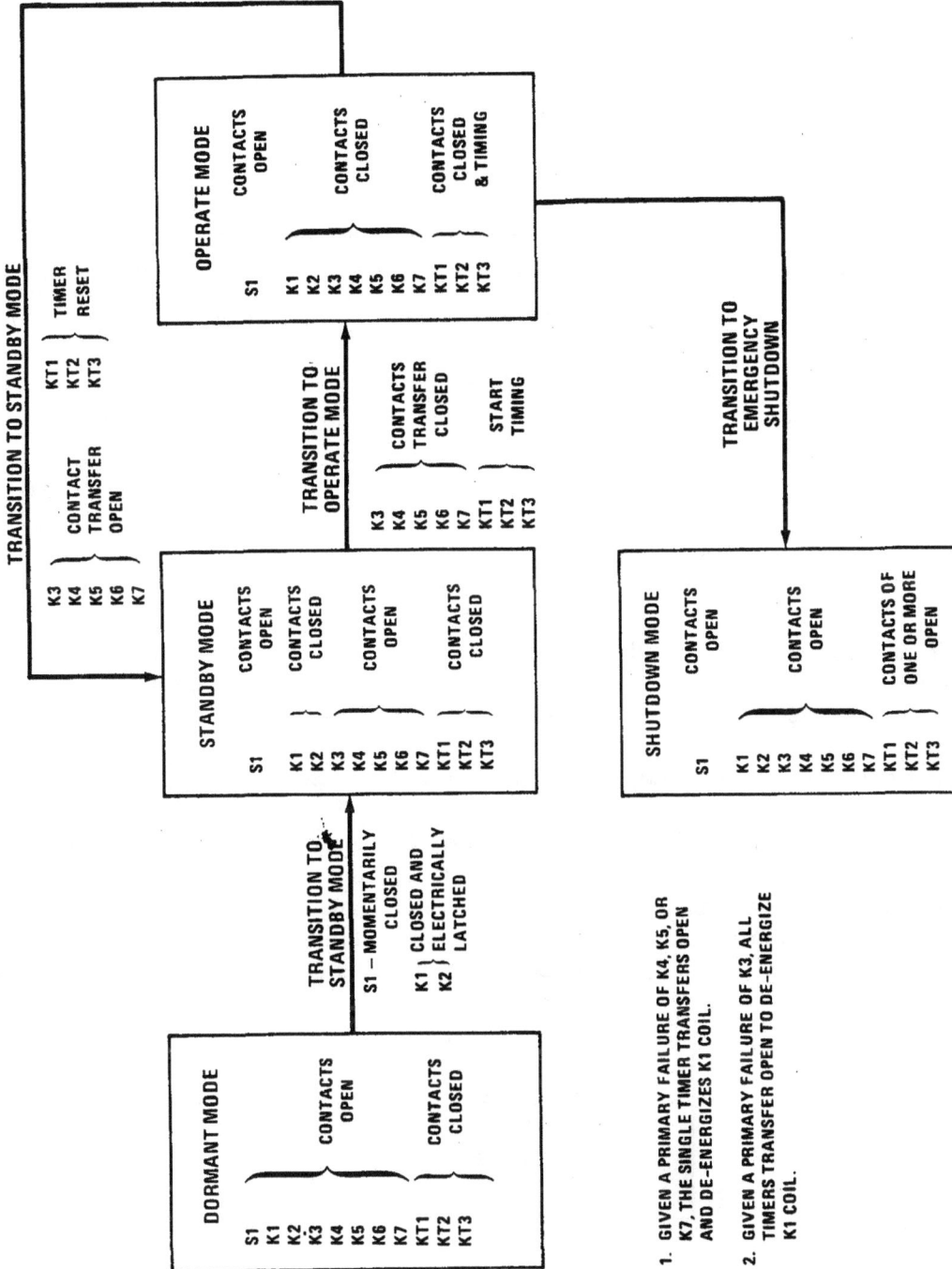

Figure IX-2. Power Distribution Box Component Status Diagram

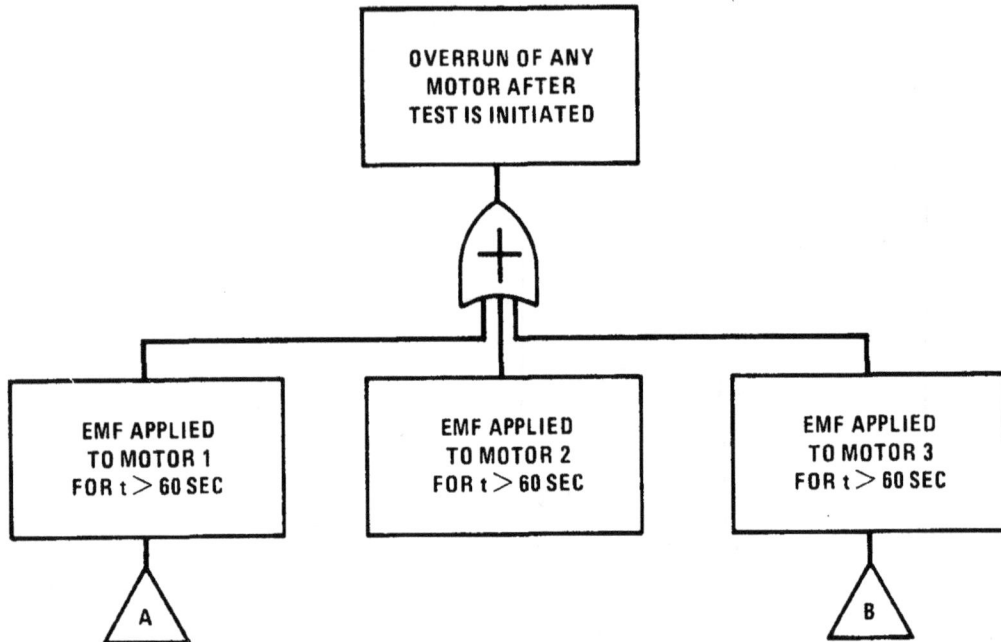

Figure IX-3. "Treetop" for Motor Overrun Problem

Our main concern in this problem is the "overrun" (t > 60 seconds) of any one of the motors after test initiation. Thus our "tree top" appears as in Figure IX-3.

Each of the three inputs to the OR-gate must be analyzed separately. We propose to treat the overrun of Motor 2 first. To avoid a clutter of trees on a single sheet, "transfer out" symbols, as shown, are applied to the fault events relevant to Motor 1 and Motor 3.

At this point, we have to decide how exhaustive our analysis is going to be. The simplest approach would be to restrict ourselves to the failures of relays and switches, and we shall do this first. A fuller analysis (which we shall hold in abeyance for awhile) would also include possible wiring defects. Limiting the analysis to the former problem for the moment, we are faced with the top event:

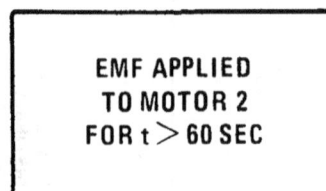

Because this is a state-of-system fault, we look for immediate necessary and sufficient causes. We see from Figure IX-1 that two fault events have to exist to cause the occurrence of the event of interest:

(1) K5 relay contacts remain closed for t > 60 seconds.

(2) K2 relay contacts fail to open when K5 relay contacts have been closed for t > 60 seconds.

These two fault events are shown as inputs to the 2nd level AND-gate in Figure IX-4 which displays the completed tree for overrun of Motor 2. Notice that the failures of K5 or K2 contacts to open are not, of themselves, faults; they become faults, however, when the time restriction is specified.

We now direct our attention to the fault event:

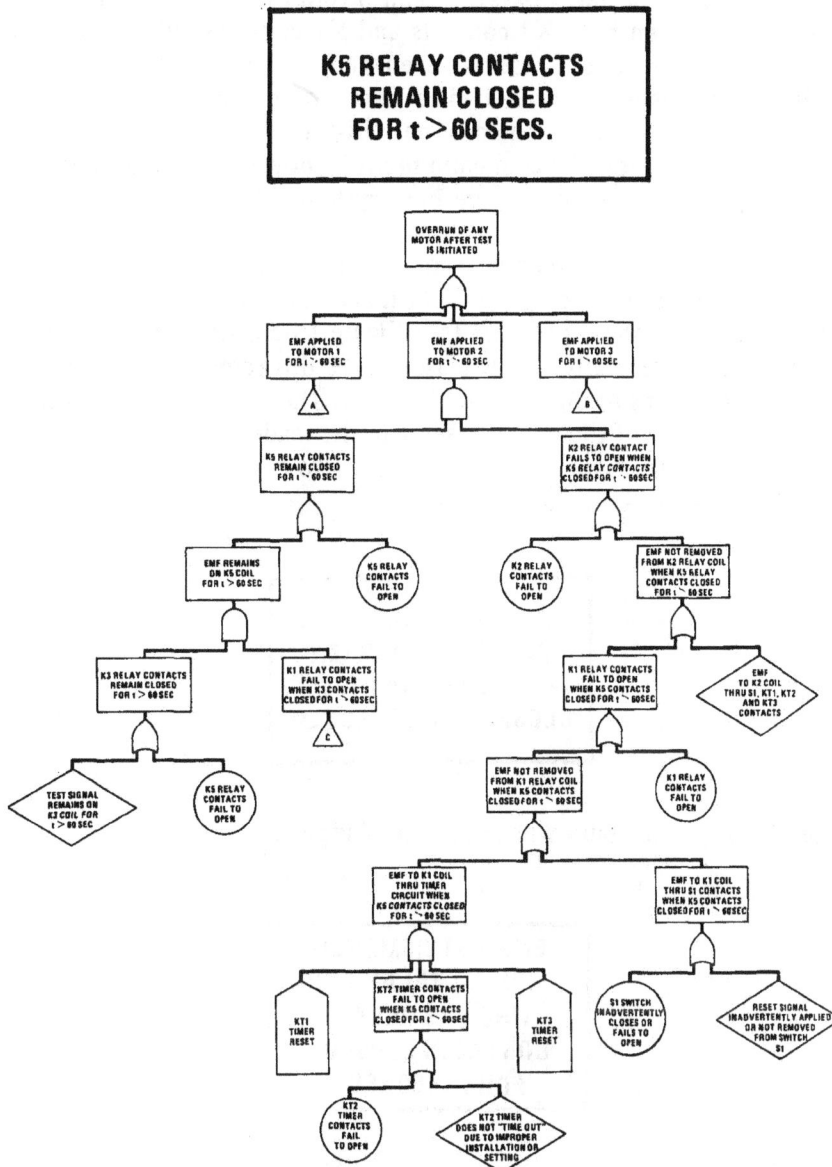

Figure IX-4. Fault Tree for Overrun of Motor 2 (Relay Logic Only)

Because this is a state-of-component fault, it demands an OR-gate with primary, secondary, and command inputs. To simplify the problem, let us not pursue secondary failures. Thus, Figure IX-4 shows only primary and command inputs. The primary failure of K5 relay is already a basic tree input (a limit of resolution); we turn our attention therefore, to the command input which is

```
┌─────────────────────┐
│                     │
│    EMF REMAINS      │
│    ON K5 COIL       │
│   FOR t > 60 SEC    │
│                     │
└─────────────────────┘
```

We note that this event is a state-of-system fault. From Figure IX-1 we find that this fault event will occur when both K3 contacts and K1 contacts fail to open with the time restriction t > 60 seconds. These two events are shown in Figure IX-4 as inputs to the 4th level AND-gate on the left. The fault event relevant to relay K3 is a state-of-component fault which demands an OR-gate followed by primary and command inputs. In this case, the command input is shown in a diamond because its cause (or causes) lie "outside" of our defined system. This leg of the tree is now complete.

Returning now to the fault event relevant to K1 relay, we realize that it will be the "top event" of a fairly substantial subsidiary tree of its own. We therefore "transfer out" to another sheet of paper (see Figure IX-5). As a matter of fact, this strategy proves to be particularly useful inasmuch as this same fault event recurs in the analysis of the overruns of Motor 1 and Motor 3. The task of studying the details of the subsidiary fault tree of Figure IX-5 is left to the reader.

Referring once more to Figure IX-4, we now have to consider the righthand input to the 2nd level AND-gate. This fault event is

```
┌─────────────────────┐
│                     │
│  K2 RELAY CONTACTS  │
│  FAIL TO OPEN WHEN  │
│  K5 RELAY CONTACTS  │
│ CLOSED FOR t > 60 SEC │
│                     │
└─────────────────────┘
```

This is a state-of-component fault whose command input is

```
┌─────────────────────┐
│                     │
│  EMF NOT REMOVED    │
│ FROM K2 RELAY COIL  │
│   WHEN K5 RELAY     │
│  CONTACTS CLOSED    │
│   FOR t > 60 SEC    │
│                     │
└─────────────────────┘
```

a state-of-system fault. This is really a single input event except that we have also chosen to consider the possibility (highly improbable) of an EMF on the K2 coil through the S1, KT1, KT2 and KT3 contacts. This latter event appears in a diamond in Figure IX-4.

The details of the rest of the right leg of the tree are left to the reader. The basic inputs are primary failures of K1 relay, S1 switch and KT2 timer contacts.

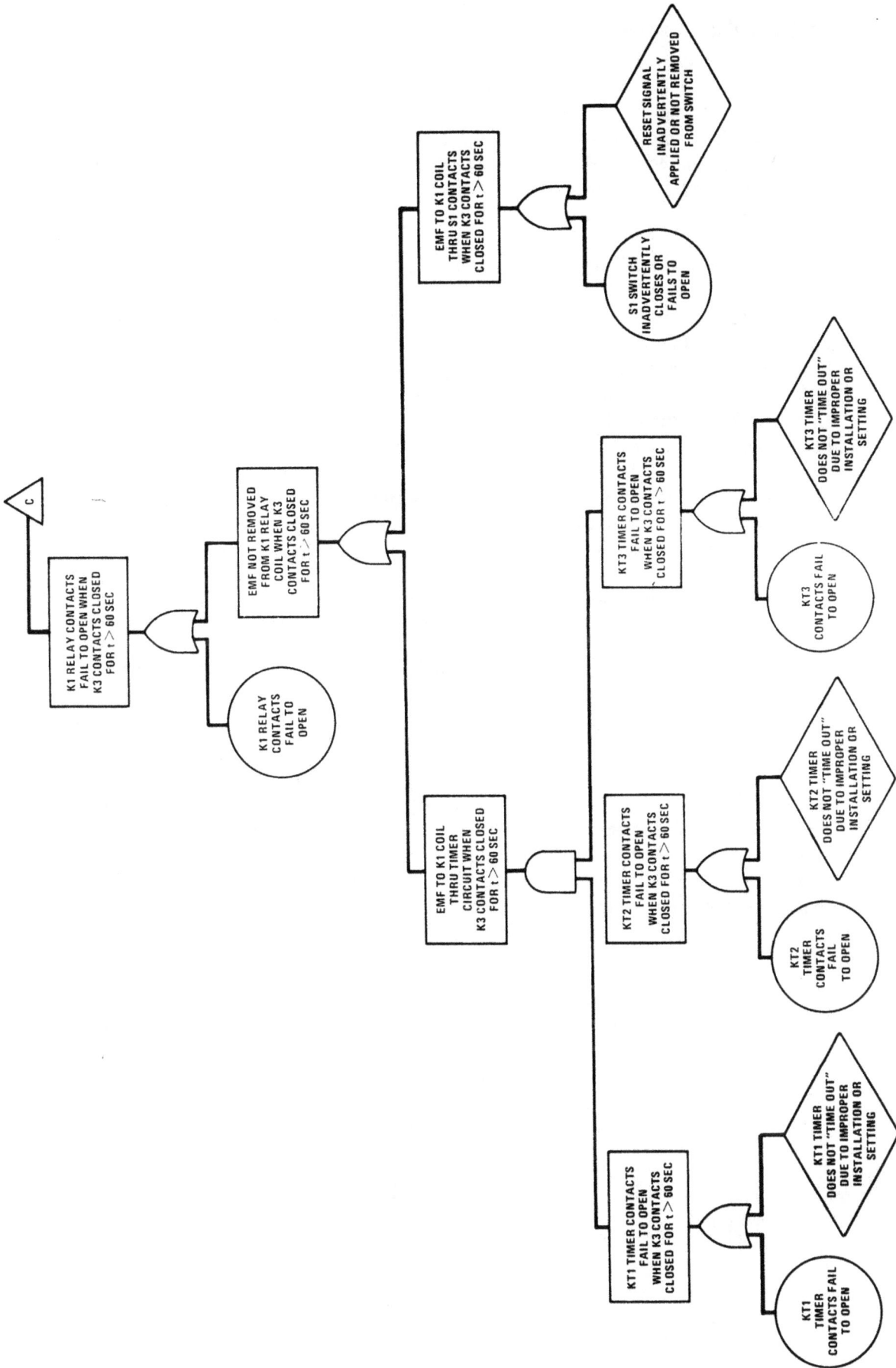

Figure IX-5. Analysis of Fault Event Relevant to K1 Relay

The fault trees relevant to overruns of Motor 1 and Motor 3 are shown in Figures IX-6 and IX-7. A more exhaustive study of the overrun problem for Motor 2 (including wiring defects) is shown in Figures IX-8, and IX-9. The reader may now care to try his hand at the analysis of a different top event: Motor 1 fails to start in presence of test signal. A solution for Motor 1 is displayed in Figure IX-10.

2.   Fault Tree Evaluation (Minimal Cut Sets)

We now turn to the qualitative evaluation of the fault tree of Figure IX-4 and determine the minimal cut sets for the overrun of Motor 2. For ease of presentation we will deal with a reduced version of the fault tree of Figure IX-4 where the diamonds and houses are removed. The reduced tree is shown in Figure IX-11.

In Figure IX-11 the circles represent primary failures of the components denoted, and the fault events E_1, E_2, etc. are defined as follows:

E_1 —   EMF applied to Motor 2 for t > 60 seconds.

E_2 —   K5 relay contacts remain closed for t > 60 seconds.

E_3 —   K2 relay contacts fail to open when K5 contacts have been closed to t > 60 seconds.

E_4 —   EMF remains on K5 coil for t > 60 seconds.

E_5 —   K1 relay contacts fail to open when K5 contacts have been closed for t > 60 seconds.

E_6 —   K1 relay contacts fail to open when K3 contacts have been closed for t > 60 seconds.

E_7 —   EMF not removed from K1 relay coil when K3 contacts have been closed for t > 60 seconds.

The reader should note that events E_5 and E_6, although similar, are not the same because the "when" is different. Thus, the tendency to transfer the fault event E_5 to the subsidiary tree of Figure IX-5 must be strongly resisted.

In the equations below each fault event is expressed in terms of its equivalent Boolean equation. Engineering notation is used.

$$E_1 = E_2 \cdot E_3$$
$$E_2 = K5 + E_4$$
$$E_3 = K2 + E_5$$
$$E_4 = K3 \cdot E_6$$
$$E_5 = K1 + KT2 + S1$$
$$E_6 = K1 + E_7 + S1$$
$$E_7 = KT1 \cdot KT2 \cdot KT3$$

Next, starting from the bottom of the tree, each fault event is expressed in terms of the basic tree inputs.

$$E_7 = KT1 \cdot KT2 \cdot KT3$$
$$E_6 = K1 + (KT1 \cdot KT2 \cdot KT3) + S1$$
$$E_5 = K1 + KT2 + S1$$

$$E_4 = K3 \cdot (K1 + KT1 \cdot KT2 \cdot KT3 + S1)$$
$$\quad = K3 \cdot K1 + K3 \cdot (KT1 \cdot KT2 \cdot KT3) + K3 \cdot S1$$
$$E_3 = K2 + K1 + KT2 + S1$$
$$E_2 = K5 + K3 \cdot K1 + K5 + K3 \cdot (KT1 \cdot KT2 \cdot KT3) + K5 + K3 \cdot S1$$
$$E_1 = [K5 + K3 \cdot K1 + K5 + K3 \cdot (KT1 \cdot KT2 \cdot KT3) + K5 + K3 \cdot S1]$$
$$\quad \cdot [K2 + K1 + KT2 + S1]$$

The Boolean expression for E_1 simplifies to:

$$E_1 = \{K5 + (K3 \cdot K1) + [K3 \cdot (KT1 \cdot KT2 \cdot KT3)] + (K3 \cdot S1)\}$$
$$\quad \cdot \{K2 + K1 + KT2 + S1\}$$

Expanding this expression using the distributive law, the minimal cut sets (all redundancies eliminated) are:

$$K5 \cdot K2$$
$$K5 \cdot K1$$
$$K5 \cdot KT2$$
$$K5 \cdot S1$$
$$K3 \cdot K1$$
$$K3 \cdot S1$$
$$K3 \cdot KT1 \cdot KT2 \cdot KT3$$

Note, for instance, that because $(K3 \cdot K1)$ is minimal from the relationship $(K3 \cdot K1 \cdot K1) = (K3 \cdot K1)$, then $(K3 \cdot K1 \cdot K2)$, for example, is not minimal.

All the minimal cut sets above are "doubles" except the last one which appears to be a "quadruple." If all three timers, however, are identical, they might be likely candidates for a so-called "common case" failure. In this case, our ostensible quadruple minimal cut set becomes essentially a "double." In later sections, we will further discuss common cause failures.

The reader should now try his hand at identifying the minimal cut sets in the other two problems: overrun of Motor 1 and overrun of Motor 3. Following the procedure detailed for Motor 2 he should be able to check the following solutions:

MINIMAL CUT SETS	MINIMAL CUT SETS
(Overrun of Motor 1)	(Overrun of Motor 3)
$K1 \cdot K4$	$K1 \cdot K7$
$K4 \cdot S1$	$K2 \cdot K7$
$K4 \cdot KT1$	$K7 \cdot KT3$
$K1 \cdot K3$	$K7 \cdot S1$
$K3 \cdot S1$	$K1 \cdot K3$
$K3 \cdot KT1 \cdot KT2 \cdot KT3$	$K3 \cdot S1$
	$K3 \cdot KT1 \cdot KT2 \cdot KT3$

THREE MOTOR EXAMPLE

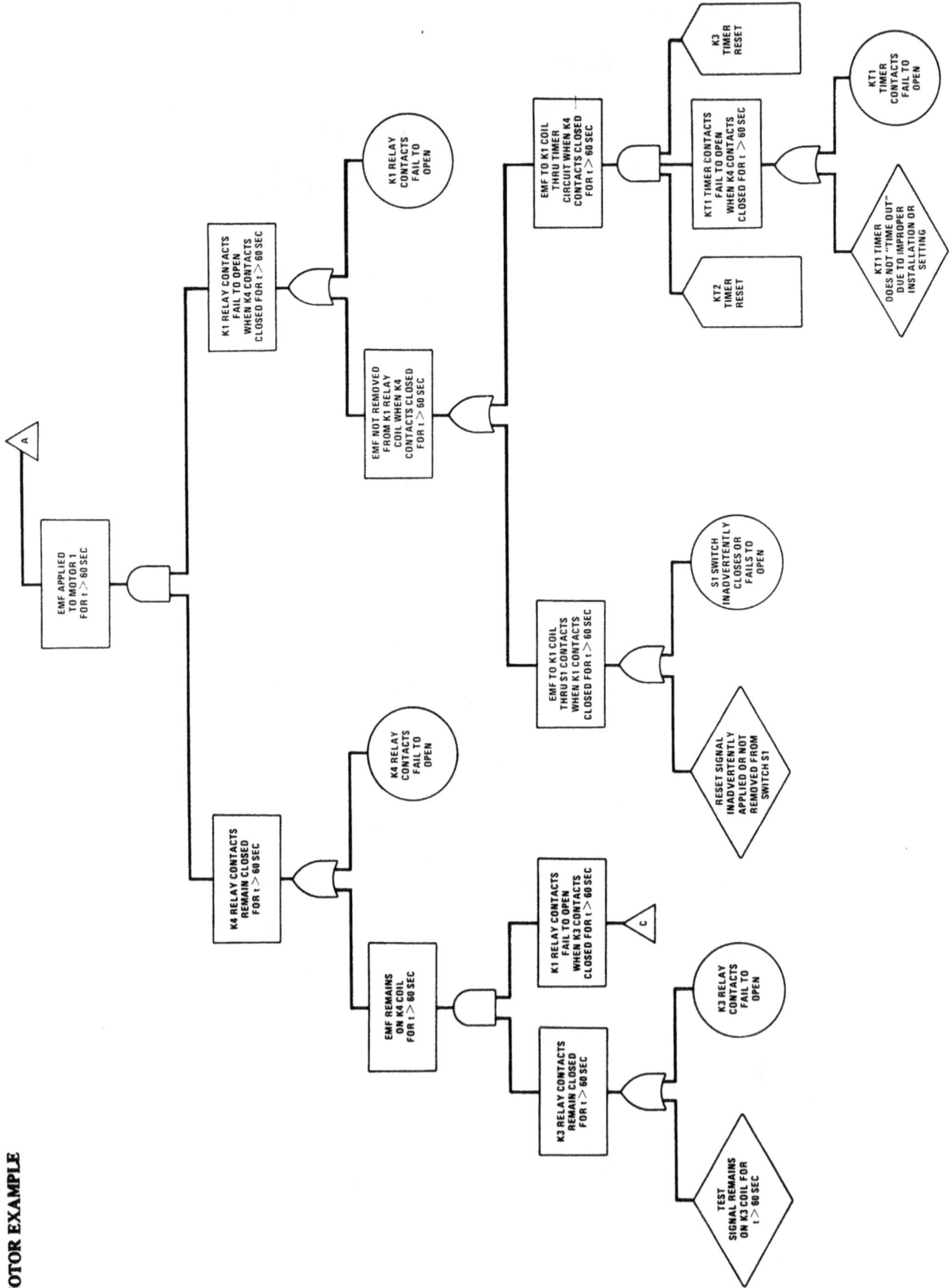

Figure IX-6. Fault Tree for Overrun of Motor 1 (Relay Logic Only)

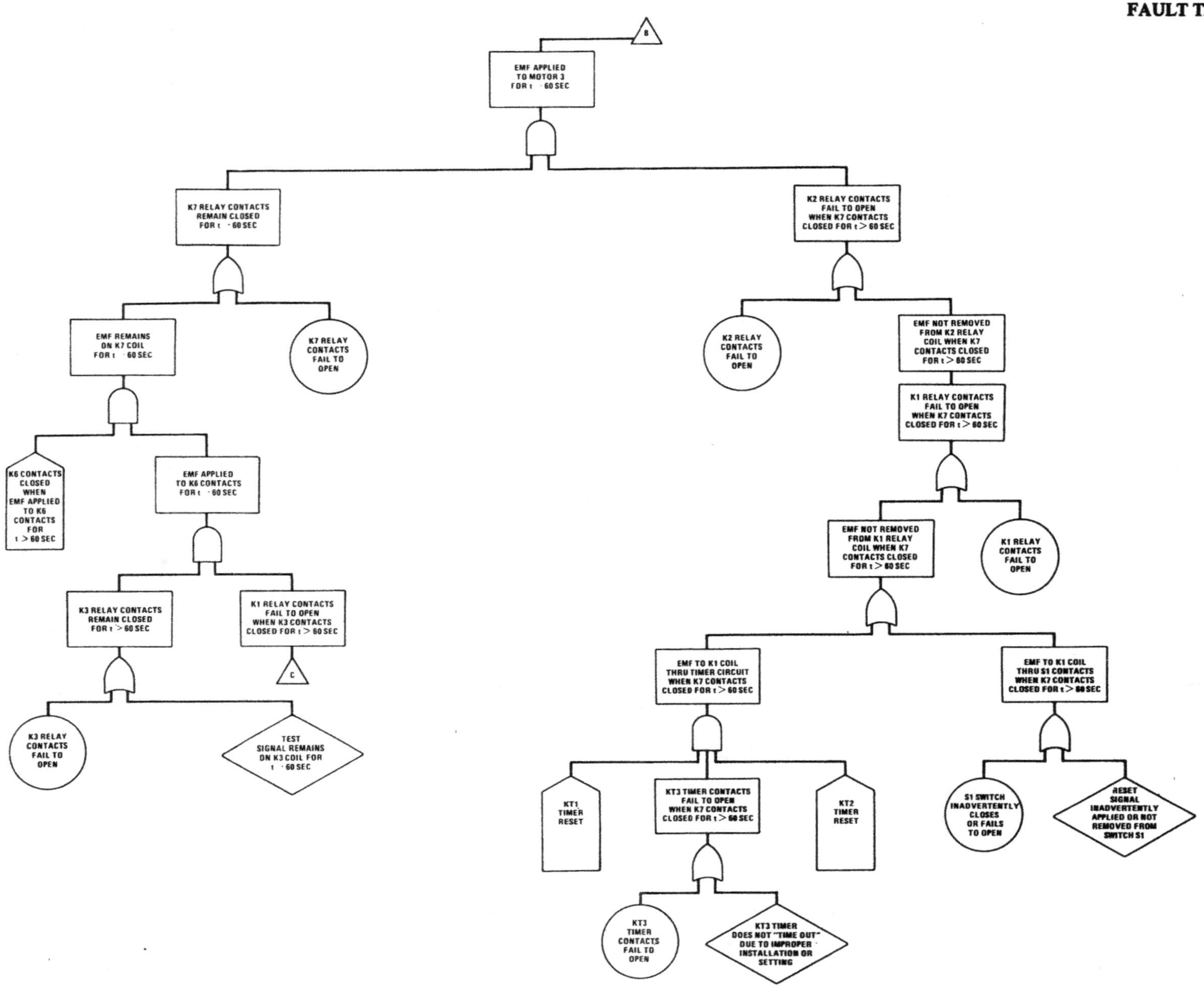

Figure IX-7. Fault Tree for Overrun of Motor 3 (Relay Logic Only)

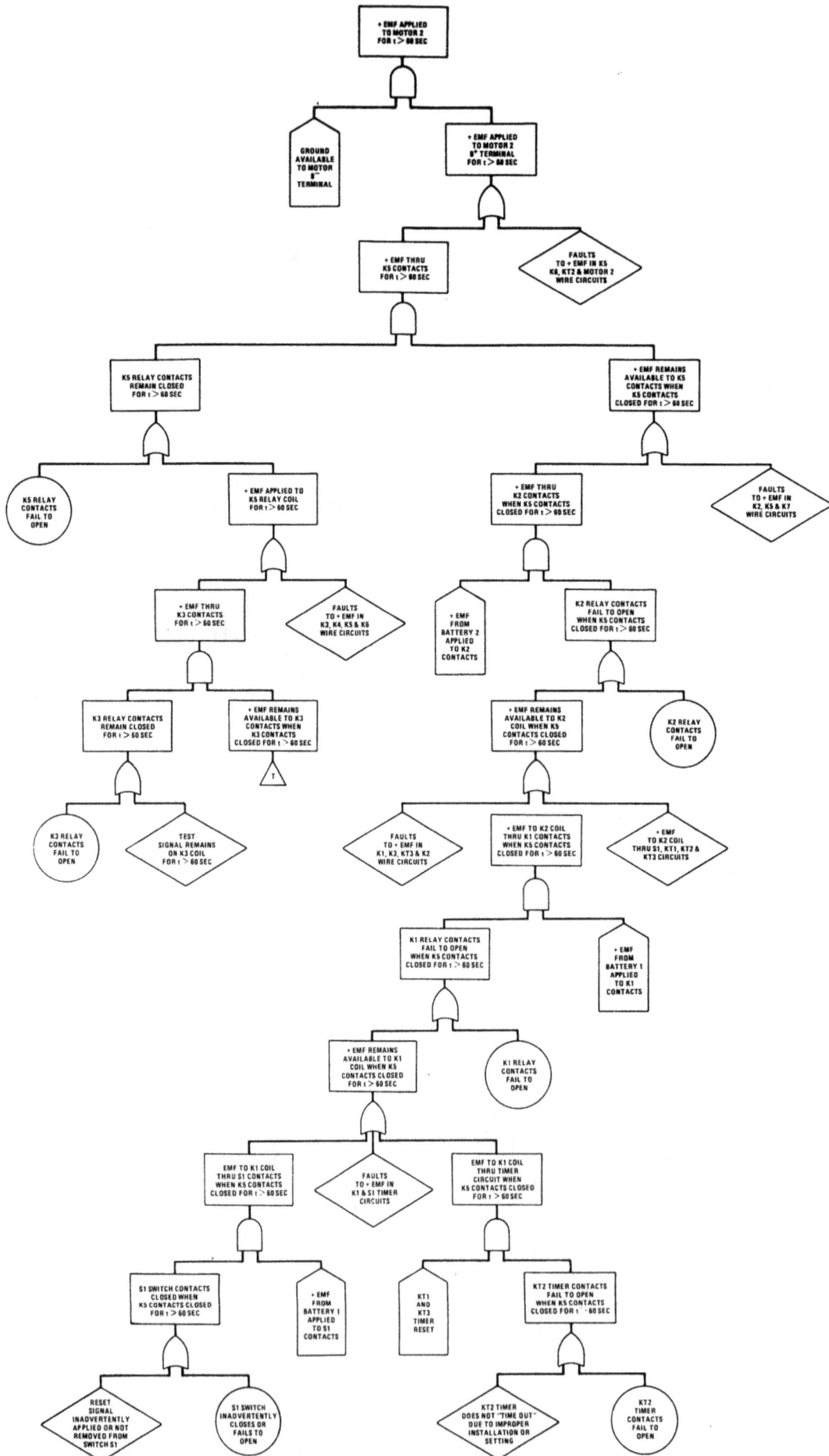

Figure IX-8. Fault Tree for Overrun of Motor 2 (Including Wiring Defects)

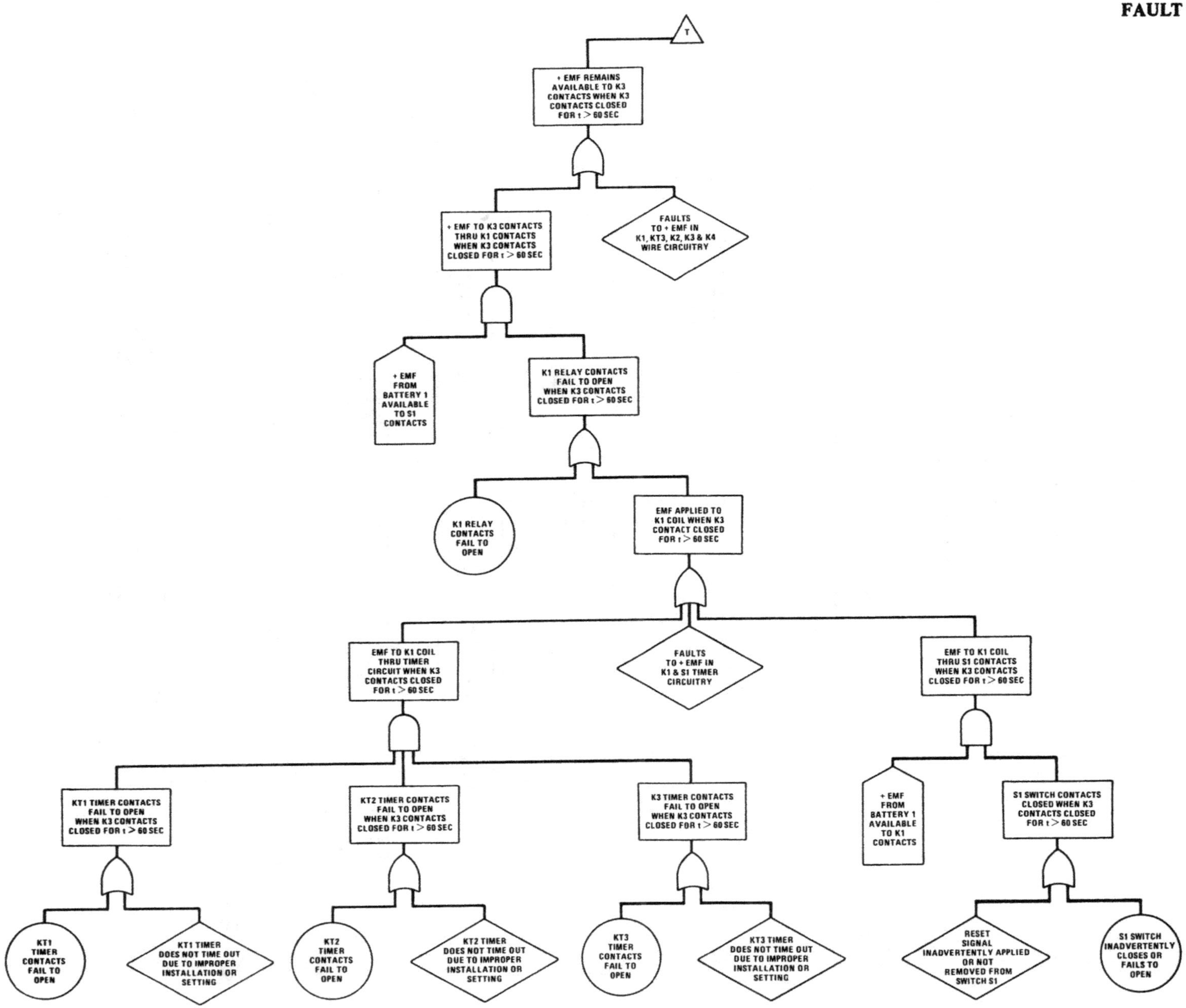

Figure IX-9. Subtree for Transfer "T" in Motor Overrun Problem

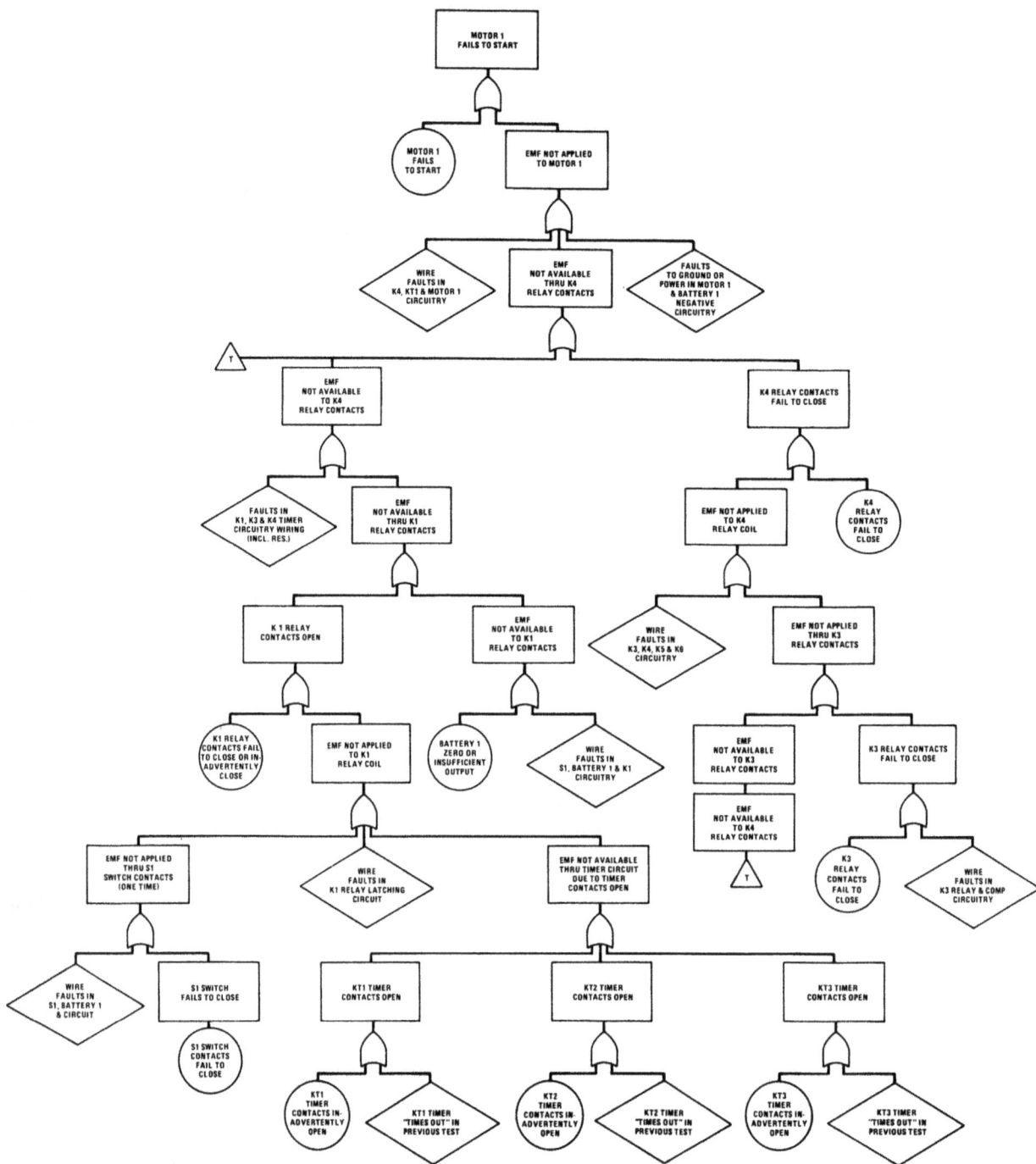

Figure IX-10. Fault Tree for Motor 1 Failure to Start

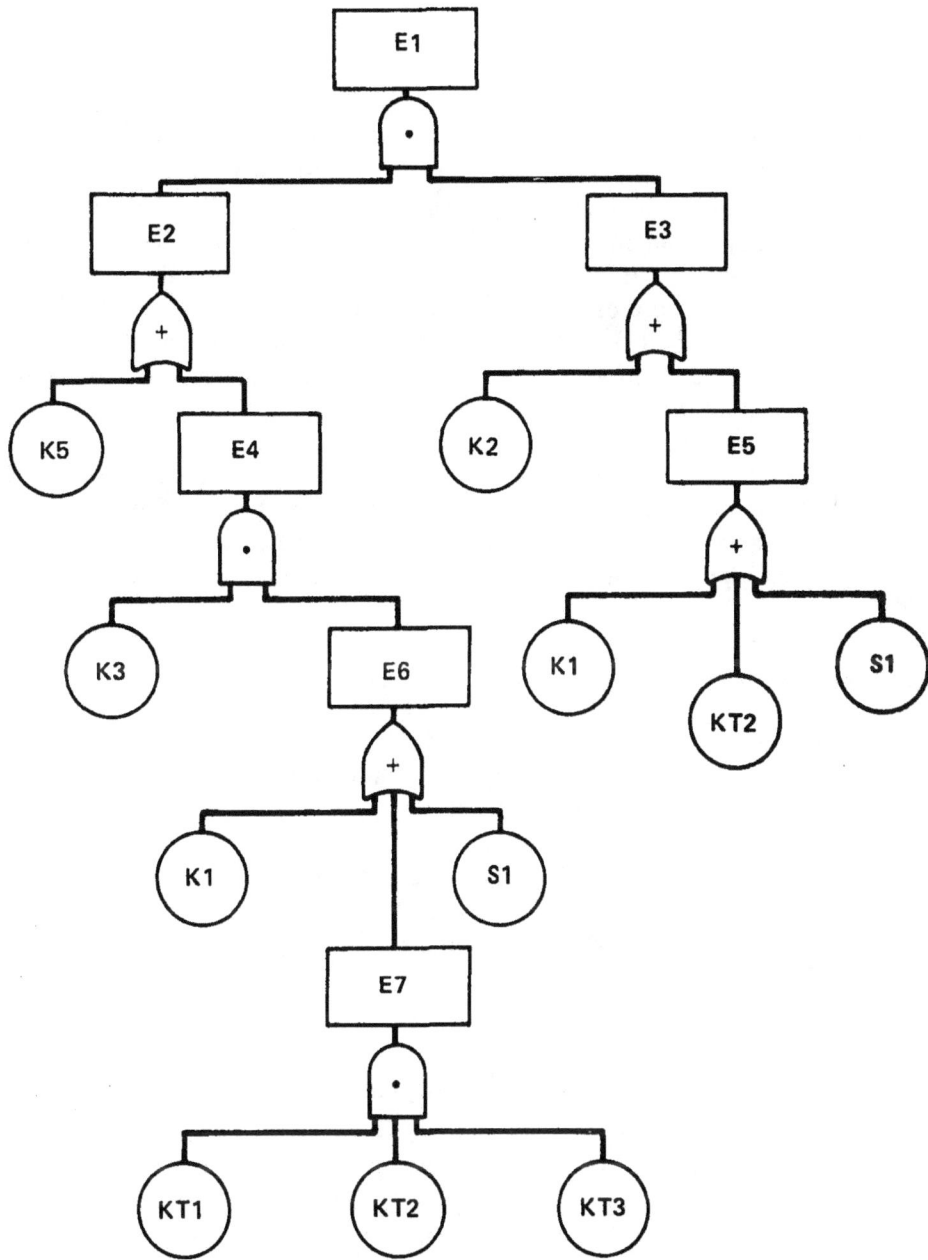

Figure IX-11. Basic Fault Tree for Overrun of Motor 2

CHAPTER X – PROBABILISTIC AND STATISTICAl ANALYSES

1. Introduction

In this chapter we shall attempt to present those basic elements that are necessary for understanding the probabilistic and statistical concepts associated with the fault tree. This material will also serve as the basis for future chapters on the quantification of fault trees. The reader who feels well-versed in statistics and probability may prefer to skip this chapter and proceed directly to Chapter XI. He may find it convenient at a later time to review this chapter as the need arises.

We shall start with a discussion of probability distribution theory. To introduce this discussion we shall first look at the binomial distribution and shall then proceed to distribution theory in general, with emphasis on some specific distributions that are of special interest to the systems analyst. We shall then attack the basics of statistical estimation.

The method of presentation, although perhaps not in the best tradition of mathematical statistics, has been developed by one of the authors in the course of teaching statistics to engineering students and engineers in the field. In this process mathematical rigor has sometimes been sacrificed to expedite the explication of basic ideas.

2. The Binomial Distribution

Suppose that we have 4 similar systems that are all tested for a specific operating time. At the end of each test let us further suppose that the test results allow us to categorize each run unambiguously as "success" or "failure." If the probability of success on each run is designated by p (and thus the probability of failure is $(1-p)$), what is the probability of $x = 0, 1, 2, 3, 4$ successes out of the four runs or trials?

The outcome collection for this experiment can be easily written down (subscripts denote run number, and "S" and "F" stand for success and failure):

$\underline{x = 4}$	$\underline{x = 3}$	$\underline{x = 2}$	$\underline{x = 1}$	$\underline{x = 0}$
		$S_1 S_2 F_3 F_4$		
	$S_1 S_2 S_3 F_4$	$S_1 F_2 S_3 F_4$	$F_1 F_2 F_3 S_4$	
$S_1 S_2 S_3 S_4$	$S_1 S_2 F_3 S_4$	$F_1 S_2 S_3 F_4$	$F_1 F_2 S_3 F_4$	$F_1 F_2 F_3 F_4$
	$S_1 F_2 S_3 S_4$	$F_1 S_2 F_3 S_4$	$F_1 S_2 F_3 F_4$	
		$S_1 F_2 F_3 S_4$		
	$F_1 S_2 S_3 S_4$	$F_1 F_2 S_3 S_4$	$S_1 F_2 F_3 F_4$	

A symbol such as $S_1 F_2 S_3 F_4$ means, "first run successful, second run a failure, third run successful, fourth run a failure." The probability of this particular outcome is $p \cdot (1-p) \cdot p \cdot (1-p) = p^2 \cdot (1-p)^2$. Note that in 4 trials, it is possible to have 4 successes (x = 4) in only 1 way, 3 successes (x = 3) in 4 ways, 2 success (x = 2) in 6 ways, 1 success (x = 1) in 4 ways, and no successes (x = 0) in only 1 way. In general for n trials the number of ways of achieving x successes is

$$\binom{n}{x} = \frac{n!}{x! \, (n-x)!}$$

which is the number of combinations of n items taken x at a time. Note also that there are a total of $16 = 2^4$ outcomes. If there were n trials and each trial could be classified as "success" or "failure," there would be 2^n outcomes.

Now let us classify these results in a somewhat different way as follows:

No. of Successes	No. of Ways	Probability
x = 0	1	$1p^0(1-p)^4$
x = 1	4	$4p^1(1-p)^3$
x = 2	6	$6p^2(1-p)^2$
x = 3	4	$4p^3(1-p)^1$
x = 4	1	$1p^4(1-p)^0$

Consider the terms in the last column. For example, $4p^3(1-p)$ represents the probability of having exactly 3 successes in 4 trials. Three successes in 4 trials can occur in 4 ways (thus the coefficient 4) and if we have had 3 successes, we must have had also 1 failure. The probability of a particular outcome of three successes and one failure is $p^3(1-p)$ and because there are 4 possible outcomes in which three successes are obtained the total probability is $4p^3(1-p)$. These expressions in the last column represent individual terms in the <u>binomial distribution</u>, the general form of which is written as follows:

If the probability of success on any trial is p, then

$$P[\text{exactly x successes in n trials}] = \binom{n}{x} p^x (1-p)^{n-x}$$

$$\equiv b\,(x; n, p)$$

(X-1)

in which $b\,(x; n, p)$ represents the binomial distribution in what is termed <u>probability density form</u>. (The probability density form is discussed further in succeeding sections.)* If the reader will set n = 4 and let x range through the values of 0 to 4, he will see that the individual terms in the probability column of our previous example are generated by equation (X-1).

_____

*For a discrete variable such as x, the probability density is also sometimes called the probability mass function.

Questions regarding the probabilities of having at most x successes in n trials and at least x successes in n trials can be answered by summing the appropriate individual terms. Thus:

$$P[\text{at most x successes in n trials}] = \sum_{s=0}^{x} \binom{n}{s} p^s(1-p)^{n-s} \equiv B(x; n,p) \qquad (X\text{-}2)$$

$$P[\text{at least x successes in n trials}] = \sum_{s=x}^{n} \binom{n}{s} p^s(1-p)^{n-s}$$

$$= 1 - \sum_{s=0}^{x-1} \binom{n}{s} p^s(1-p)^{n-s} \qquad (X\text{-}3)$$

where B $(x; n,p)$ is the binomial distribution in cumulative distribution form (the cumulative distribution form is discussed further in upcoming sections). At this stage we may simply interpret the cumulative binomial as giving the probability that the number of successes is less than or equal to some value. Thus, returning to our example, the probability of having at least 2 successes in 4 trials is,

$$6p^2(1-p)^2 + 4p^3(1-p) + p^4 = 1 - \left[(1-p)^4 + 4p(1-p)^3\right].$$

The binomial distribution is extensively tabulated, often in the form of equation (X-2) but also occasionally in the form of equation (X-3) and sometimes in the form of equation (X-1). See for example References [1], [33], and [41].

The statistical average of the binomial distribution is np and its variance is $np(1-p)$. The average is a measure of the location of the distribution and the variance is a measure of its dispersion, or spread. These terms are discussed further in subsequent sections.

We have made a number of assumptions (tacitly so far) in our own use of the binomial. It is of the utmost importance to list these assumptions explicitly:

(1) Each trial has only one of two possible outcomes. We have called these "Success"/"Failure" but they could just as well be called anything else (e.g., defective/nondefective).

(2) There are exactly n random trials and n is numerically specified.

(3) All n trials are mutually independent.

(4) The probability of "success" (or whatever you want to call it) on any trial is designated by some letter such as p, and p remains constant throughout the sequence of trials.

It is important for the reader to realize that if a problem comes up and one or more of these assumptions is violated, then use of the binomial distribution will be questionable unless the effect of the violation is investigated. All distributions and, as a matter of fact, all mathematical formulae are characterized by underlying assumptions and restrictions, and a valid use of these distributions or formulae involves a knowledge of what the associated assumptions and restrictions are.

Let us now return to our list of assumptions and consider them in more detail. What can be done in case one or more of these restrictions happens not to be true? We will look at only a few cases to indicate the possible violations to the assumptions.

(1) What if each trial has more than 2 possible outcomes? For instance, in certain testing methods, 3 decisions are possible after each trial: accept the lot, reject the lot, continue testing. If we are drawing single chips from a container holding a mixture of white, green, red, blue, and yellow chips, then each trial has 5 possible outcomes. This case does not pose a serious problem, for, instead of using the binomial distribution, we simply use its generalization, which is known as the multinomial distribution and which is adequately discussed in many statistics texts (see, e.g., reference [52]). When applicable, we might also lump the outcomes into "success" and "failure" and use the binomial on this more gross categorization. (The probability of "success" would be the sum of the probabilities of the outcomes categorized as "success.")

(2) Now suppose that the number of trials, n, is not known, but only the number of successes. For example, we may agree to throw a single die until the number "5" comes up. It is not known, a priori, how many tosses will be necessary. Or we may agree to test similar relays until one fails. Again the number of relays tested is not known. Under such conditions (i.e., whenever a number of trials are performed until a preassigned number of successes are obtained) we cannot use the binomial distribution, but another distribution closely related to the binomial and called the negative binomial distribution is available (see reference [13]). The negative binomial $\hat{b}(x; k,p)$ gives the probability that the kth success occurs on the xth trial and is written,

$$\hat{b}(x; b,p) = \binom{x-1}{k-1} p^k (1-p)^{x-k} \qquad \text{(X-4)}$$

(3) If the outcomes of the n trials are mutually interdependent (i.e., the $(x+1)^{th}$ outcome depends in some way on the xth outcome and possibly on preceding outcomes as well), a number of difficulties arise. Some sort of conditional probability representation is required. The probability of a specific sequence of outcomes would depend on the order of occurrence ("serial outcome space") and each different sequence would have a different probability. For example, it would generally be invalid to use the binomial distribution to estimate the probability that the daily weather be rainy or not rainy because weather patterns often persist for days or even weeks at a time and what happens on Wednesday is dependent on what has happened on the preceding Tuesday. If independence is in doubt, there are statistical tests to check independence before the binomial is used (see reference [11]).

(4) If the probability of "success" (or whatever) changes throughout the series of trials, where samples are chosen from a fixed population without replacement, the so-called hypergeometric distribution can be employed (see reference [52]). This distribution is:

$$h(x; n, a, b) = \frac{\binom{a}{x}\binom{b}{n-x}}{\binom{a+b}{n}} \qquad \text{(X-5)}$$

where   a = number of items exhibiting characteristic A in the population
          b = number of items exhibiting characteristic B in the population
    a+b = N = population size or lot size
          n = size of sample drawn from the population
          x = number of items exhibiting characteristic A in the sample.

For example, characteristic A may be "defective" and characteristic B, "non-defective." Here h (x; n, a, b) gives the probability that exactly x of the n sample items exhibit characteristic A.

Use of the hypergeometric distribution is necessary when sampling of a small population is carried out without replacement. ("Small" means that N and n in equation (X-5) are comparable in size.) For instance, if we receive a shipment of 50 transistors, 10 of which are defective, the initial proportion of inoperative ones is 1/5, but this proportion will change as we withdraw a sample of, say, 20 without replacement.

As the reader will observe from equation (X-5), use of the hypergeometric frequently entails complicated arithmetical calculations involving factorial numbers, and for this reason, the binomial is often used in these cases to give approximate results. The binomial distribution provides a good approximation whenever $N \geqslant 10n$ where N is the population (or lot or batch) size and n is the sample size.* In this case a/n is taken as the approximate value of p.

As a specific example of the use of the binomial distribution, consider the following problem: ABC Corporation mass produces resistors of a certain type. Past experience indicates that one out of a hundred of these resistors is defective. Therefore, the probability of obtaining a defective resistor in a sample is p = 0.01. If a sample of 10 resistors is selected randomly from the production line, what is the probability of its containing exactly one defective resistor? We have:

$$x = 1, n = 10, p = 0.01 \text{ and}$$
$$b(x = 1; n = 10, p = 0.01) = \binom{10}{1} (0.01) (0.99)^9$$

If tables of the cumulative binomial distribution are available, this can be most readily evaluated by finding $B(1) - B(0)$ because

$$B(1) = P[0 \text{ or } 1 \text{ defective resistors}], \text{ and}$$
$$B(0) = P[\text{exactly } 0 \text{ defective resistors}].$$

$$B(1) - B(0) = 0.9957 - 0.9044 = 0.0913$$

By making calculations similar to the foregoing we can now draw a distribution function giving the probability of finding x = 0, 1, 2, 3, . . . , 10 defective resistors in the sample of 10. This distribution is shown in Figure X-1. The continuous curve in Figure X-1 is not really legitimate because the binomial distribution is inherently discrete, but this kind of continuous interpolation serves to make the general shape of the distribution visible.

---

*Some authors recommend a somewhat less conservative rule of thumb of $N \geqslant 8n$.

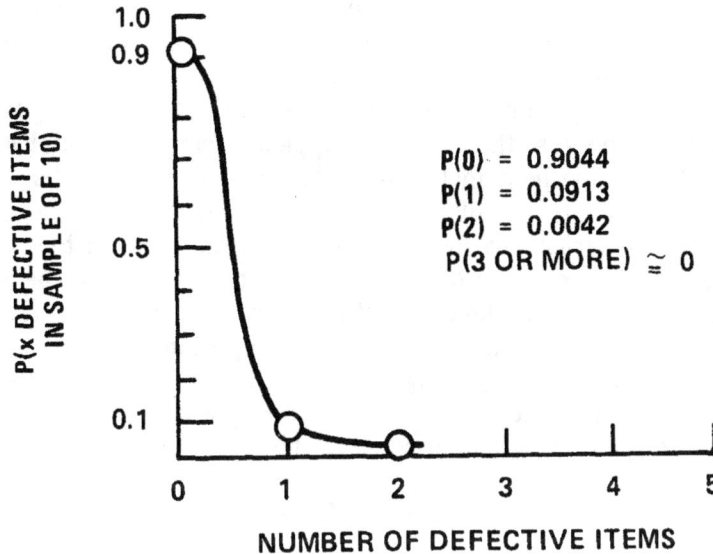

Figure X-1. Distribution of x Defective Items in a Sample of 10
when p = 0.01

In safety and reliability evaluations, the binomial distribution is applicable to systems featuring redundancy if each redundant component operates independently and if each redundant component has the same (or approximately the same) probability of failure. For example, suppose we have n redundant components and, if more than x of these fail, the system fails. Let p be the probability of component failure in some designated operating period. The probability that the system will not fail is then the probability that x or fewer components fail and this is just the binomial cumulative probability $B(x; n, p)$.

Alternatively, consider a situation in which n events are possible and, if more than x of these events occur, we have an accident. If the n events are independent and if they all have approximately the same probability of occurring, the binomial distribution is again applicable. In general, the binomial distribution may be used when we have n repetitions of some event ("n trials") and we desire information about the probability of x occurrences of an outcome, fewer than x occurrences of an outcome, or more than x occurrences of an outcome, always presuming that the assumptions previously listed are satisfied. The "n trials" can be n components, n years, n systems, n human actions, or any other appropriate quantity.

We shall return to the binomial distribution shortly because two of its limiting forms are of special interest to us. For the moment, however, we will discuss distributions and distribution parameters in general.

3. The Cumulative Distribution Function

Let us use the symbol **X** to designate the possible results of a random experiment. **X** is usually referred to as a <u>random variable</u>* and it may take on values that are either discrete (e.g., the number of defective items in a lot) or continuous (e.g., the heights or weights of a population of men). Actually, the latter category of values is also strictly discrete because the measuring apparatus employed will have some limit of resolution. It is convenient mathematically, however, to consider such values as representing a continuous variable. It will be convenient to use the corresponding lower case letter x to designate a specific value of the random variable.

The fundamental formula that we are about to present will be given for the continuous case with detours here and there to point out differences between the continuous and discrete cases whenever it seems important to do so. In general, it is operationally a matter of replacing integral signs with summation signs. The <u>cumulative distribution function</u> $F(x)$ is defined as the probability that the random variable **X** assumes values less than or equal to the specific value x.

$$F(x) = P[\mathbf{X} \leqslant x] \qquad\qquad (X\text{-}6)$$

According to equation (X-6), $F(x)$ is a probability and thus must assume values only between (and including) zero and one:

$$0 \leqslant F(x) \leqslant 1$$

If X ranges from $-\infty$ to $+\infty$, then

$$F(-\infty) = 0$$
$$F(+\infty) = 1$$

If X has a more restricted range, $x_1 < \mathbf{X} < x_u$, then

$$F(x_1) = 0$$
$$F(x_u) = 1$$

An important property of the cumulative distribution function is that $F(x)$ never decreases in the direction of increasing x. $F(x)$ is a non-decreasing function although not necessarily monotonically so, in the strict mathematical sense. This can be stated more succinctly as follows:

$$\text{If } x_2 > x_1, \text{ then } F(x_2) \geqslant F(x_1)$$

One further important property of $F(x)$ is stated in equation (X-7) below.**

$$P[x_1 \leqslant \mathbf{X} \leqslant x_2] = F(x_2) - F(x_1) \qquad\qquad (X\text{-}7)$$

---

*Random variables will be denoted in this text by boldface type.
**For discrete variables the formula is $P[x_1 < \mathbf{X} \leqslant x_2] = F(x_2) - F(x_1)$.

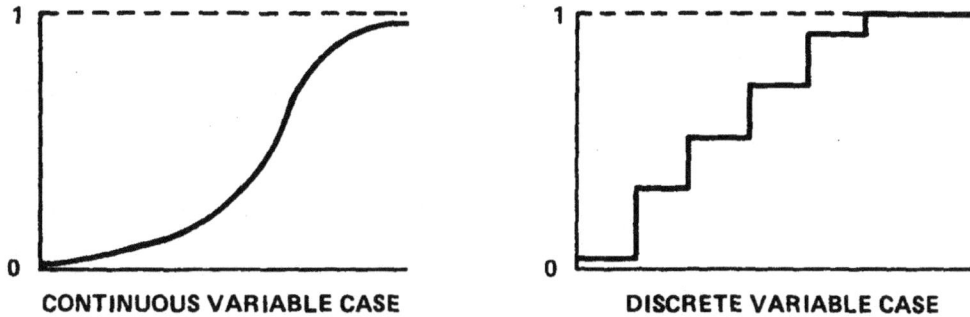

Figure X-2. Typical Shapes for F(x)

The binomial cumulative distribution $B(x; n, p)$ encountered in section 2 is one specific example of $F(x)$. Typical shapes for $F(x)$ for a continuous variable and for a discrete variable are shown in Figure X-2.

The properties of the cumulative distribution function that we have presented in the above formulae are generally valid for continuous and discrete random variables.

As an example of a random variable and its corresponding cumulative distribution, consider a random experiment in which we observe times to failure of a single component. Whenever the component fails, we repair it, set $t = 0$ and note the new time to failure (see sketch below).

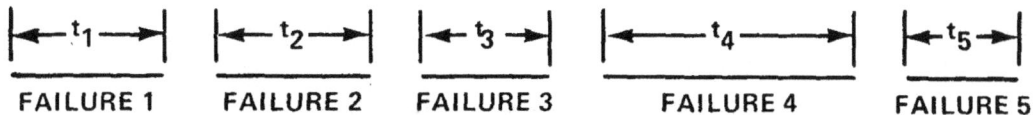

Let us assume that repair ("renewal") does not alter the component, i.e., in every instance its repair state coincides with its initial operating state. The random variable of interest, T, is the time to failure from renewal or repair. We represent a specific value of T by the symbol t_i. The cumulative distribution $F(t)$ for any t thus gives the probability that the time to failure will be less than or equal to t.

As another example consider a random experiment in which we are performing repeated measurements on some item. The random variable X represents the "measurement outcome" in general and x_i represents some specific measurement value. The cumulative distribution $F(x)$ gives the probability that the measurement value is less than or equal to x. We could estimate values of $F(x)$ from $F_{est}(x_i)$ where

$$F_{est}(x_i) = \frac{n_i}{n}$$

in which n gives the total number of measurements and n_i is the number of measurements in which X assumes values less than or equal to x_i. As n gets larger and larger $F_{est}(x_i)$ will approach more and more closely the true value $F(x_i)$. In application, the cumulative distribution function must either be determined from theoretical considerations or be estimated by statistical methods.

4. The Probability Density Function

For a continuous random variable, the <u>probability density function</u> (pdf), f(x), is obtained from F(x) by a process of differentiation:*

$$f(x) = \frac{d}{dx} F(x) \qquad \text{(X-8)}$$

An equivalent statement is,

$$F(x) = \int_{-\infty}^{x} f(y)dy \qquad \text{(X-9)}$$

Because f(x) is defined as the slope of a non-decreasing function, we must have

$$f(x) \geqslant 0 \qquad \text{(X-10)}^{\cdot}$$

When the pdf is integrated over the entire range of its argument, the result is unity.

$$\int_{-\infty}^{\infty} f(x)dx = 1 \qquad \text{(X-11)}$$

This property of f(x) permits us to treat areas under f(x) as probabilities.
The fundamental meaning of the pdf is stated in equation (X-12).

$$f(x)dx = P[x < \mathbf{X} < x + dx] \qquad \text{(X-12)}$$

Our previous equation (X-7) can now be stated in another, especially useful, form:

$$P[x_1 \leqslant \mathbf{X} \leqslant x_2] = \int_{x_1}^{x_2} f(x)dx \qquad \text{(X-13)}$$

Typical shapes for f(x) are illustrated in Figure X-3 in which (a) represents a symmetrical distribution, (b) a distribution skewed to the right, and (c) a distribution skewed to the left. (In the figures x increases as we move to the right).
In the case of a continuous variable, probabilities must be expressed in terms of intervals. This is because the probability associated with some specific value x is always zero because there are an infinite number of values of $\mathbf{X}$ in any range. Thus f(x)dx is the probability that the quantity of interest will lie in the interval between x and x + dx. Of course, the interval length dx may be made as small as we please. The quantity f(x) itself is therefore the probability per unit interval. In the case of

---

*We assume F(x) is well-behaved and allows differentiation.

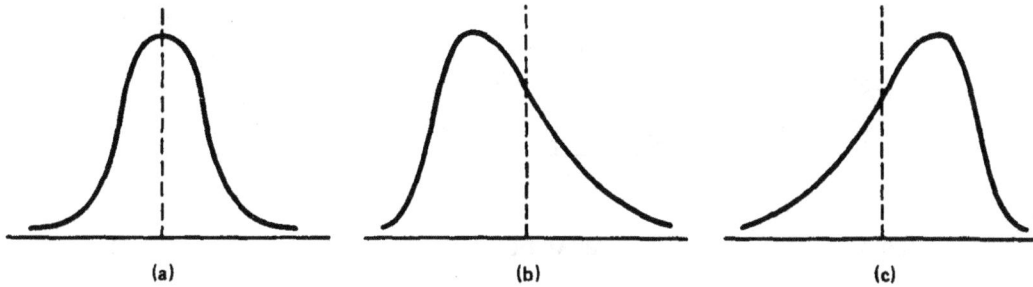

Figure X-3. Typical Shapes for f(x)

discrete variables we replace integral signs by summation signs and sum over the individual probabilities of the x-values in the range of interest. Equation (X-13), for example, then becomes applicable for a discrete variable **X**.

With regard to our previous time to failure example, f(t)dt gives the probability that the component will fail in the time interval between t and t + dt. In the measurement example f(x)dx gives the probability that the measurement outcome will lie between x and x + dx. Empirically, if we consider a large number of measurements, f(x)dx could be estimated by

$$f(x)\Delta x = \frac{\Delta n_i}{n}$$

where n is the total number of measurements and Δn_i is the number of measurements for which **X** lies between x and x + Δx.

5. Distribution Parameters and Moments

The characteristics of particular probability density functions are described by distribution parameters. One variety of parameter serves to locate the distribution along the horizontal axis. For this reason such a parameter is called a location parameter.

The most common location parameter is the statistical average. Other location parameters commonly employed are: the median (50% of the area under the probability density curve lies to the left of the median; the other 50%, to the right), the mode, which locates the "peak" or maximum of the probability curve (there may be no "peak" at all or there may be more than one as in bimodal or trimodal distributions); the mid-range, which is simply the average of the minimum and maximum values when the variable has a limited range, and others of lesser importance. For an illustration of these concepts refer to Figure X-4.

In (a) the median is indicated by $x_{.50}$. From the definition of the median, 50% of the time the outcome will be less than or equal to $x_{.50}$, and 50% of the time it will be greater than $x_{.50}$. Therefore $P(X \leqslant x_{.50}) = .50$ and, in terms of the cumulative distribution, $F(x_{.50}) = .50$. The median is a particular case of the general α-percentile, x_α, defined such that $F(x_\alpha) = \alpha$. For example, the 90% percentile is such that $F(x_{.90}) = .90$ and 90% of the time the outcome value x will be less than or equal to $x_{.90}$.

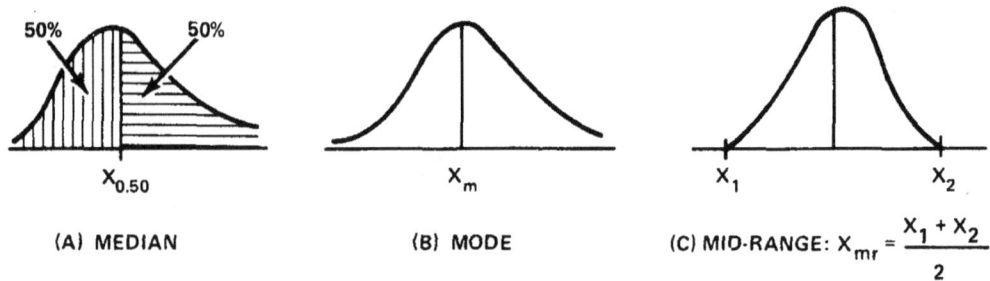

Figure X-4. The Median, Mode, and Mid-Range

In (b) of Figure X-4, the mode is indicated by x_m and gives the most probable outcome value. In (c), we see how the mid-range is calculated from the two extreme values x_1 and x_2.

The average is also termed the mean, or the expected value. If we repeat a random experiment many times and average our outcome values, this empirical average will approximate the true average and will approach the true average more and more closely as the number of repetitions is increased. (We assume the distribution has an average and is such that the empirical average converges to the population average.)

In the case of a symmetrical distribution as shown in (a) of Figure X-3, the mean, median, and mode all coincide. For skewed distributions, as in Figure X-3 (c), the median will fall between the mode and the mean. In Figure X-5 we see two symmetrical distributions with the same values of mean, median, and mode. They are, however, strikingly different from the standpoint of how the values are clustered about the central position. Parameters used to describe this aspect of a distribution are known as dispersion parameters. Of these, the most familiar are the variance and the square root of the variance or standard deviation. Other dispersion parameters, less frequently employed, are the median absolute deviation and the range between a lower bound value and an upper bound value. We will calculate that variance of the distribution in a later section.

There are, in fact, many other types of distribution parameters, but those we have mentioned are some of the most basic ones. It is essential that we become familiar with the general methods for calculating distribution parameters once the functional

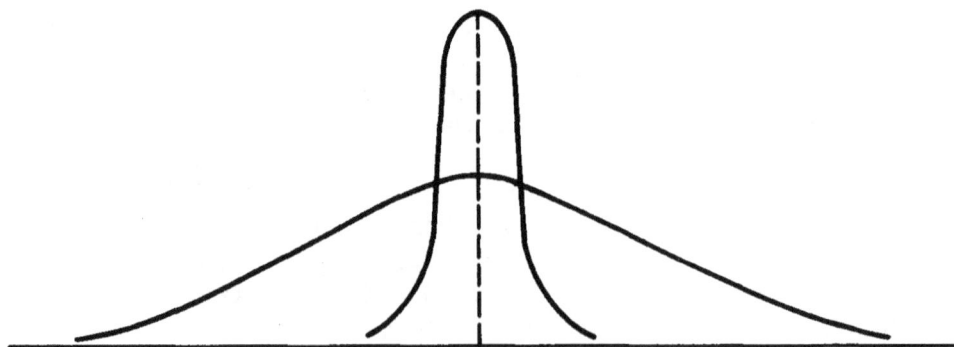

Figure X-5. Two Symmetrical Distributions with
Different Dispersion Parameters

form of the pdf is known. Some of these general methods entail calculating the
moments of the distribution and are of great importance in theoretical statistics.
Distribution moments may be calculated about any specified point but we shall
restrict ourselves to the discussion of (a) moments about the origin, and (b) moments
about the mean.

(a) Moments About the Origin

The first moment about the origin is defined as

$$\mu_1' = \int_{-\infty}^{\infty} x\ f(x)dx \qquad\qquad (X\text{-}14)$$

This gives us the mean or expected value of X, written $E[X]$. We will use the symbol
μ for the mean for the sake of brevity although actually $E[X] = \mu$.

The second moment about the origin is defined as

$$\mu_2' = \int_{-\infty}^{\infty} x^2\ f(x)dx \qquad\qquad (X\text{-}15)$$

and gives us the expected value of X^2, $E[X^2]$.

In general, the nth moment about the origin is defined as

$$\mu_n' = \int_{-\infty}^{\infty} x^n\ f(x)dx \qquad\qquad (X\text{-}16)$$

and gives us the expected value of X^n, $E[X^n]$.

If $Y = g(X)$ is any function of X, and X is distributed according to the pdf $f(x)$,
the expected value of $g(X)$ may be obtained from

$$E[Y] = E\Big[g(X)\Big] = \int_{-\infty}^{\infty} g(x)\ f(x)dx \qquad\qquad (X\text{-}17)$$

(b) Moments About the Mean

The first moment about the mean is defined as

$$\mu_1 = \int_{-\infty}^{\infty} (x-\mu)\ f(x)dx \qquad\qquad (X\text{-}18)$$

Because μ_1 is always and invariably equal to 0, it is not of great utility.

The second moment about the mean is defined as

$$\mu_2 = \int_{-\infty}^{\infty} (x-\mu)^2 \, f(x)dx. \tag{X-19}$$

This gives us the variance σ^2 or $E\left[(X-\mu)^2\right]$. In general, the nth moment about the mean is defined as

$$\mu_n = \int_{-\infty}^{\infty} (x-\mu)^n \, f(x)dx \tag{X-20}$$

and gives us $E\left[(X-\mu)^n\right]$.

There is a useful relationship between μ_2, μ_2', and μ_1', namely,

$$\mu_2 = \mu_2' - (\mu_1')^2. \tag{X-21}$$

Equation (X-21) permits us to calculate the variance by evaluating the integral in (X-15) rather than the integral in (X-19) which is more complicated algebraically. Equation (X-21) is easily proved as follows:

$$\mu_2 = \int_{-\infty}^{\infty} (x-\mu)^2 \, f(x)dx$$

$$= \int_{-\infty}^{\infty} x^2 \, f(x)dx - 2\mu \int_{-\infty}^{\infty} x \, f(x)dx + \mu^2 \int_{-\infty}^{\infty} f(x)dx$$

$$= \mu_2' - 2\mu^2 + \mu^2 = \mu_2' - \mu^2 = \mu_2' - (\mu_1')^2.$$

In the case of a <u>discrete</u> random variable, the first moment about the origin is written

$$\mu = \mu_1' = \sum_{i=1}^{n} x_i \, p(x_i) \tag{X-22}$$

where $p(x_i)$ is the probability associated with the value x_i. The commonly used formula for finding the average of n values,

$$\bar{x} = \frac{1}{n} \sum_{i=1}^{n} x_i,$$

is a special case of equation (X-22) which can be used whenever all the individual values are treated as having the same probability or "weight," namely 1/n. Thus, in the case of a single die we have

$$\mu = \mu' = \frac{1+2+3+4+5+6}{6} = \frac{21}{6} = 3.5$$

so that the expected value is 3.5 despite the fact that such an outcome is impossible in practice.

Also, if the random variable is discrete, the second moment about the mean (the variance) assumes the form:

$$\mu_2 = \sum_{i=1}^{n} (x_i - \mu)^2 \, p(x_i) \qquad (X\text{-}23)$$

In case all the x_i have the same "weight" $\frac{1}{n}$, equation (X-23) reduces to a sampling formula for computing the variance of a sample of n readings,*

$$s^2 = \frac{1}{n} \sum_{i=1}^{n} (x_i - \overline{x})^2. \qquad (X\text{-}24)$$

We conclude this section with a simple example of the use of distribution moments. Consider the rectangular pdf shown in Figure X-6 in which any value between a and b is equally likely. Because any value is equally likely, $f(x) = f_0$, a constant. The area under a pdf must integrate to 1 and hence we have

$$\text{Area} = f_0 \, (b-a) = 1 \text{ so that } f_0 = \frac{1}{b-a}.$$

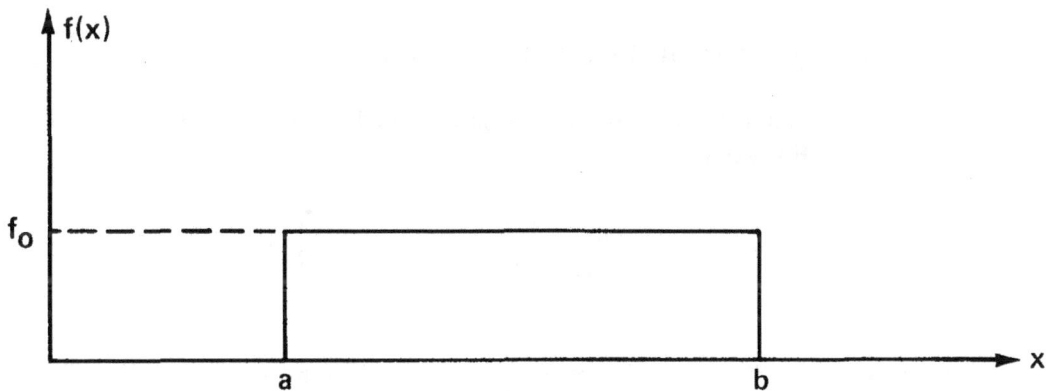

Figure X-6. The Rectangular Distribution

---

*(X-23) provides a biased estimate of the population variance σ^2. The bias is eliminated by multiplying (X-23) by the factor n/(n–1), "Bessel's correction." For large n the difference is slight. The question of bias is discussed later in this chapter.

The mean of this distribution is given by

$$\mu = \mu_1' = E(X) = \int_a^b x \, \frac{1}{b-a} \, dx$$

$$= \frac{1}{b-a} \left[\frac{x^2}{2} \right]_a^b = \frac{b^2 - a^2}{2(b-a)} = \frac{(b-a)(b+a)}{2(b-a)} = \frac{b+a}{2} \, .$$

The variance of the distribution is given by

$$\mathrm{Var} = \sigma^2 = \mu_2 = \mu_2' - (\mu_1')^2 = \int_a^b x^2 \left(\frac{1}{b-a} \right) dx - \left(\frac{b+a}{2} \right)^2$$

$$= \left(\frac{1}{b-a} \right) \left[\frac{x^3}{3} \right]_a^b - \left(\frac{b+a}{2} \right)^2$$

$$= \left(\frac{1}{b-a} \right) \left(\frac{b^3 - a^3}{3} \right) - \left(\frac{b+a}{4} \right)^2$$

$$= \left(\frac{1}{b-a} \right) \left[\frac{(b-a)(b^2 + ab + a^2)}{3} \right] - \left(\frac{b+a}{4} \right)^2$$

$$= \frac{b^2 - 2ab + a^2}{12} = \frac{(b-a)^2}{12} \, .$$

6. Limiting Forms of the Binomial: Normal, Poisson

Several important distributions can be obtained as limiting forms of the binomial distribution. For example, consider

$$\lim_{\substack{p \text{ fixed} \\ n \to \infty}} \left[\binom{n}{x} p^x (1-p)^{n-x} \right].$$

We read the above as the limit when p is fixed and n goes to infinity. Omitting the mathematical details, this process leads to the famous normal or Gaussian* distribution

$$f(x; \mu, \sigma) = \frac{1}{\sqrt{2\pi} \, \sigma} \exp \left[-\frac{1}{2} \left(\frac{x-\mu}{\sigma} \right)^2 \right] \qquad \text{(X-25)}$$

---

*We suspect that the normal distribution is sometimes called Gaussian in commemoration of the fact that it was one of the few things Gauss <u>didn't</u> discover! Actually it was first investigated by de Moivre.

where μ and σ are the mean and standard deviation respectively. The normal distribution is extensively tabulated but not in the form (X-25). A direct tabulation of (X-25) would require an extensive coverage of values of the parameters μ and σ and would be impossibly unwieldly. It is possible, however, to find a transformation which has the effect of standardizing μ to 0 and σ to 1. This transformation is

$$z = \frac{x - \mu}{\sigma}$$ (X-26)

and the corresponding distribution in terms of z is

$$f(z) = \frac{1}{\sqrt{2\pi}} e^{-z^2/2}$$ (X-27)

which is known as the standardized normal distribution and which forms the basis for all tabulations.

The reader should note that the passage from equation (X-25) to equation (X-27) via equation (X-26) is not a matter of simple substitution. The transformation from one distribution to another requires, among other things, a factor known as the Jacobian of the transformation (see reference [25]). In the present case the Jacobian is just σ, which neatly cancels the σ in the original coefficient $1/(\sqrt{2\pi}\,\sigma)$. A graph of $f(z)$ is shown in Figure X-7.

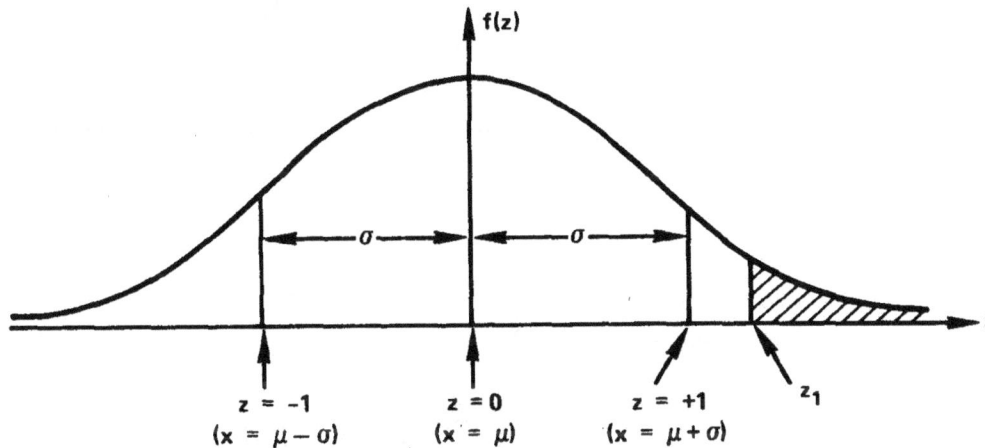

Figure X-7. The Standardized Normal Distribution

Several characteristics of the standardized normal distribution are tabulated. Of primary interest to us is the area under the curve included between two points on the horizontal axis. The reader should remember that such an area can be treated as a probability because the total area under the curve is unity. Suppose that an arbitrary point z_1 is a point of interest (see Figure X-7). Some tabulations record the area under the curve from z_1 to $+\infty$ (shaded in the figure). Some tabulations record the area from z_1 to $-\infty$ and still others, the area from z_1 to the origin. It is, of course, essential to ascertain what area is being tabulated before the tables can be used intelligently.

For the normal distribution, in terms of the original variable **X**, σ measures the distance from the mean μ to the inflection point of the curve. For **X**, the area under the pdf curve from $\mu - \sigma$ to $\mu + \sigma$ is approximately 0.68 and the area from $\mu - 2\sigma$ to $\mu + 2\sigma$ is approximately 0.95.

It is assumed that the reader already has some familiarity with the normal distribution and its tabulations. Nonetheless, a simple numerical example will be given, which can be skipped by the experienced. The widths of slots in forgings are normally distributed with mean $\mu = 0.9000$ inch and standard deviation $\sigma = 0.0030$ inch. If the specification limits (acceptance limits) are 0.9000 inch $\pm$ 0.0050, what percentage of the total output will be rejected? The rejected forgings will be those widths that fall into either one of the shaded regions indicated in the sketch.

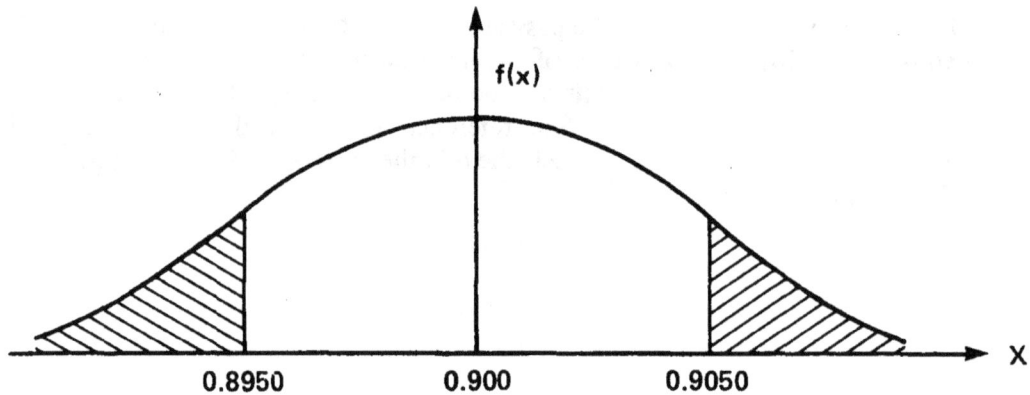

The value of z corresponding to x = 0.9050 is

$$z = \frac{0.9050 - 0.9000}{0.0030} = 1.67$$

From standard normal tables, the right-hand tail area for $P[Z \geqslant 1.67]$ is 0.0475. This is the probability that $\mathbf{X} \geqslant 0.9050$. By symmetry the left-hand tail area is also 0.0475. Then the area of both tails is 2(0.0475) = 0.0950. This is the probability of finding a slot width that is outside the specifications. Therefore, 9.5% of the total production will be rejected.

This rejection rate is fairly high. If no change in the specifications is possible, we could lower the rejection rate by controlling production more carefully so that σ is reduced. Suppose that our goal is a rejection rate of 1/1000 = 0.001. What is the maximum allowable σ, say σ', that is consistent with such a rejection rate?

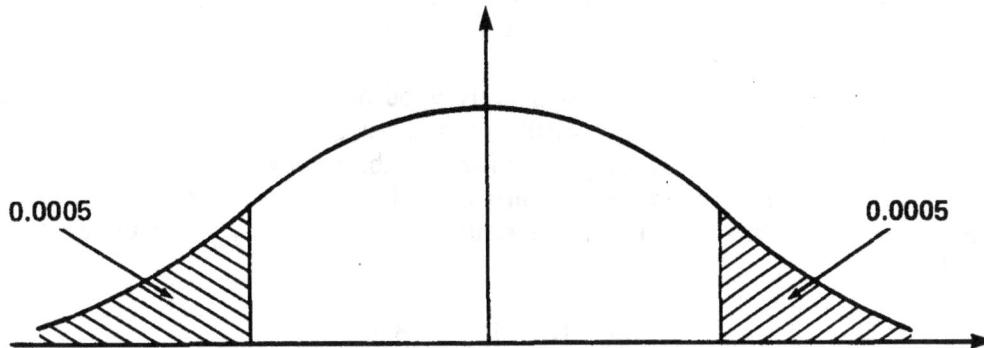

If the rejection rate is 0.001 the area in each tail (see sketch) must be 0.001/2 = 0.0005. From tables, the value of z that cuts off a tail area of 0.0005 is 3.3. From $z = (x-\mu)/\sigma$ we find $\sigma' = (x-\mu)/z$,

$$\text{thus } \sigma' = \frac{0.9050 - 0.9000}{3.3} = \frac{0.005}{3.3} = 0.00152 \text{ inch.}$$

Therefore, for a rejection rate of .001, the maximum allowable value of σ is 0.00152 inch.

There are several reasons for us to be interested in the normal distribution. One is that the mean of a large number of measurements tends to be normally distributed regardless of the underlying distribution of each measurement (from the Central Limit Theorem). Another is that the normal distribution provides a fairly good statistical model of many wear-out processes.

Another critically important distribution, from the standpoint of systems and reliability analysis, is obtained as the following limiting form of the binomial distribution:

$$\lim_{\substack{n \to \infty \\ n \to 0}} \left\{ \binom{n}{x} p^x (1-p)^{n-x} \right\} \qquad (X\text{-}28)$$

In equation (X-28) the limit is taken in such a way that the product np remains finite. The outcome of the limiting process is

$$\frac{(np)^x}{x!} e^{-np} = \frac{m^x}{m!} e^{-m} \qquad (X\text{-}29)$$

where $m = np$. (The mathematical details of the limiting process can be found in many texts, e.g., reference [32], pp. 45–46.)

Equation (X-29) gives the probability of exactly x occurrences of a <u>rare event</u> ($p \to 0$) in a large number of trials ($n \to \infty$). The expected number of occurrences of the event is np = m. The distribution given in (X-29) is known as the <u>Poisson distribution</u>. As can be shown by the method of moments (although there are easier ways) that the mean and variance of the Poisson distribution are both numerically equal to m.

The Poisson distribution provides a fairly good approximation to the binomial even though p is not particularly small and n is not especially large. For example, suppose that in a mass production process the probability of encountering a defective unit is 0.1 (p = 0.1). What is the probability of finding exactly one defective unit in a random lot of 10 (n = 10)? The exact result can be found by using the binomial distribution:

$$b(1; 10, 0.1) = 0.3874$$

The Poisson approximation yields the result 0.3679 which does not differ greatly from the exact value. If we increase the random lot size to 20 (n = 20), the agreement is even closer:

binomial − 0.2702
Poisson − 0.2707

The Poisson distribution is important not only because it approximates the binomial, but because it describes the behavior of many rare event occurrences, regardless of their underlying physical process. The Poisson distribution also has numerous applications in describing the occurrence of system (or component) failures under steady-state conditions. We will now further discuss these system applications.

7. Application of the Poisson Distribution to System Failures—The So-Called Exponential Distribution

Assume we have a given system which is in steady-state, when it is not in the process of being burnt in and is it not wearing out. Further, we will assume that when the system fails, it is restored to a condition that is essentially as good as new and that the repair time is negligible. Let the event of interest be system failure. We are specifically interested in the probability of exactly 0 occurrences of system failure. Thus, applying the Poisson distribution we are led to set x = 0 in equation (X-29). The result is:

$$P[0 \text{ occurrences of system failure}] = e^{-m}$$

where m = expected number of system failures in a large number of "trials."

Now, as far as failures are concerned, the parameter of interest is <u>time</u>. We thus seek a way of somehow expressing m in terms of time. It turns out that this can be readily done.

Assume we have data for this system and, on the average, a failure occurs every 50 hours. We say the mean time to failure (θ) is 50 hours. If we operate this system for

100 hours, we expect to experience 2 failures; that is, 100/50 = 2. Using the symbol t for operating time we have in general:

Expected number of failures in time $t = t/\theta = \lambda t$ if $\lambda = 1/\theta$

But m = expected number of failures. Therefore,

$$P[0 \text{ occurrences of system failure}] = e^{-m} = e^{t/\theta} = e^{\lambda t}.$$

Now the reliability of a system, R(t), is by definition the probability of continuous successful operation for a time t. Hence we have

$$R(t) = e^{-t/\theta} = e^{-\lambda t} \qquad (X-30)$$

The probability of system failure prior to time t is given by the cumulative distribution function F(t). The system either fails prior to time to or it does not; we have, therefore,

$$R(t) = e^{-\lambda t} = 1 - F(t)$$

and

$$F(t) = 1 - e^{-\lambda t}. \qquad (X-31)$$

The pdf corresponding to equation (X-31) can now easily be found.

$$f(t) = \frac{d}{dt} F(t) = \frac{d}{dt} (1 - e^{-\lambda t})$$

$$\qquad (X-32)$$

$$f(t) = \lambda e^{-\lambda t}$$

The pdf in equation (X-32) is generally referred to as the "exponential distribution of time-to-failure." Also, the expression (X-30) is sometimes referred to as simply the exponential distribution.

The reliability, cumulative distribution, and pdf, given by equations (X-30), (X-31), and (X-32), respectively, are widely used in system analysis and reliability. The reason is clear. The exponential is an especially simple distribution. Only one parameter (λ, the failure rate or θ, the mean time to failure) must be determined empirically. We must, however, be extremely cautious in applying equation (X-30) to the calculation of system reliability for the following reasons. We saw that equation (X-30) arises from the Poisson distribution. The latter is obtained as a limiting form of the binomial distribution. The use of the binomial distribution is restricted by a number of assumptions that we took care to list earlier in this chapter. Some of these assumptions have been modified in the limiting process, but one of them remains untouched: the assumption that all "trials" are mutually independent. Translated into failure language, a trial is an "opportunity to fail in some interval of time."

When system failures are repairable, then our interpretation of the independent trial assumption is as follows: the probability of failure in some future interval is a function of only the length of the interval and is independent of the number of past failures. If the system is not able to be repaired, our interpretation of the probability behavior must be modified to the following: given no prior failure, the probability of failure in some future time interval is a function only of the length of that interval. We need the restriction "given no prior failure" because, for a non-repairable component, if we have had a failure at an earlier time, then the probability of the failure <u>existing</u> at a later time is 1 and the probability of the failure <u>occurring</u> at a later time is 0 because it has already occurred.

Another way of characterizing the failure processes is as follows. For the exponential distribution, given no failure up to time t, the probability of failure in the interval $(t, t + \Delta t)$ is the same as the probability of failure in any interval of the same length (given no failure has occurred up to that time). Specifically, it is the same as the probability of failure in the interval $(0, \Delta t)$. Thus, because we start operation at $t = 0$, we are "as good as new" at time t, which is another description of the exponential.

If we start with the assumption that the probability of failure in a certain interval is a function only of interval length, we can derive the exponential distribution from that assumption alone. Consider, for example, a non-repairable system that can exist in one of two states: E_1—system operating, E_0—system failed. Let us define

$P_1(t)$ = probability that system is in state E_1 at time t
$P_0(t)$ = probability that system is in state E_0 at time t

and let us assume that at the start of time $(t = 0)$ the system is in its operating state E_1.

Now $P_1(t + \Delta t)$ represents the probability that the system is in state E_1 at time $t + \Delta t$. We can write

$$P_1(t + \Delta t) = P_1(t) [1 - \lambda \Delta t] = P_1(t) - P_1(t) \lambda \Delta t$$

where, according to our initial assumption (probability of failure in a given interval is a function only of interval length), the quantity $\lambda \Delta t$ gives the probability that the system makes a transition from state E_1 to state E_0 in the time interval Δt and that λ is some constant (the failure rate). Consequently $(1 - \lambda \Delta t)$ gives the probability that the system does <u>not</u> make a transition from E_1 to E_0 (i.e., does not fail) in time interval Δt. Algebraic rearrangement yields the following difference equation:

$$\frac{P_1(t + \Delta t) - P_1(t)}{\Delta t} = -\lambda P_1(t).$$

If we allow the length of the time interval to approach zero $(\Delta t \rightarrow 0)$, the limiting form of the left-hand side of the equation is just, by definition, the derivative of $P_1(t)$ with respect to t. Thus

$$\lim_{\Delta t \rightarrow 0} \left[\frac{P_1(t + \Delta t) - P_1(t)}{\Delta t} \right] = \frac{d}{dt} P_1(t) = P_1'(t) = -\lambda P_1(t)$$

in which the prime stands for differentiation with respect to time, a convention originally proposed by Newton. We now have the differential equation,

$$P_1'(t) = -\lambda P_1(t).$$

This is easily integrated if we remember our boundary condition $P_1(t=0) = 1$.

$$\frac{d[P_1(t)]}{P_1(t)} = -\lambda dt$$

$$[\ln P_1(t)]_0^t = [-\lambda t]_0^t$$

$$\ln P_1(t) - \ln 1 = -\lambda t$$

$$P_1(t) = e^{-\lambda t}$$

which is just the reliability of the system. Also, because $P_0(t) + P_1(t) = 1$, we have

$$P_0(t) = 1 - e^{-\lambda t} = 1 - R(t) = F(t).$$

The corresponding pdf is

$$f(t) = \lambda e^{-\lambda t}$$

and we recognize the exponential distribution.

8.   The Failure Rate Function

Recall from previous sections that

F(t) = P[failure occurs at some time prior to t]

and that

f(t)dt = P[failure occurs between t and t + dt].

We now define a conditional probability, $\lambda(t)$, called the <u>failure rate function</u>:

$\lambda(t)dt = P$ [failure occurs between t and t + dt | no prior failure].   (X-33)

For any general distribution, there is an important relationship between the three functions $\lambda(t)$, f(t), and F(t);

$$\lambda(t) = \frac{f(t)}{1 - F(t)}.$$ (X-34)

The validity of equation (X-34) is easily demonstrated as follows.

Let us designate the time at which failure occurs as **T**. **T** is a random variable according to the definition given in equation (X-33),

$$\lambda(t)dt = P \ [t < T < t + dt \ | \ t < T].$$

Let $(t < T < t + dt)$ be denoted as event A and $(t < T)$ be denoted as event B. We remember from basic probability theory that in general,

$$P \ (A \mid B) = \frac{P(A \cap B)}{P(B)}.$$

Therefore,

$$\lambda(t)dt = \frac{P \ [(t < T < t + dt) \cap (t < T)]}{P(t < T)}$$

Now event A is just a special case of event B, i.e., when A occurs then B automatically occurs. In set theoretic notation, $A \subset B$ (A is a subset of B) and, under these circumstances, $A \cap B = A$. It follows that

$$\lambda(t)dt = \frac{P \ [t < T < t + dt]}{P(t < T)} = \frac{P[A]}{P[B]} = \frac{f(t) \ dt}{1 - F(t)}.$$

Finally,

$$\lambda(t) = \frac{f(t)}{1 - F(t)},$$

which is just equation (X-34).

If $\lambda(t)$ is plotted against time for a general system, the curve shown in Figure X-8 results. For obvious reasons this relationship between $\lambda(t)$ and t has become known as the "bathtub curve."

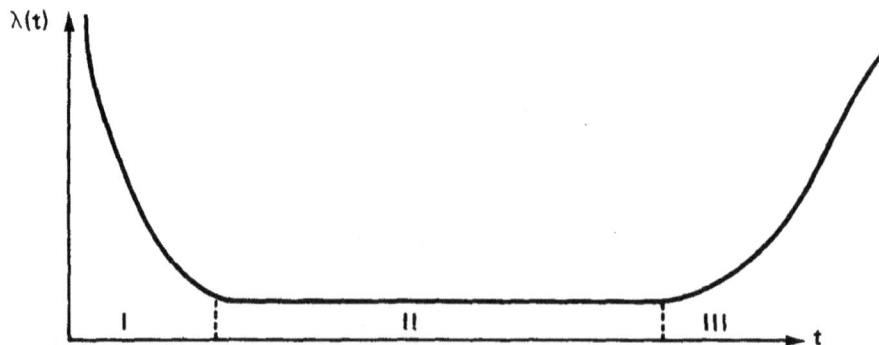

Figure X-8. Plot of $\lambda(t)$ vs. t for a General System

This curve can be divided into three parts which are labeled I, II, III in Figure X-8. Region I is termed the region of "infant mortality" where sometimes an underlying distribution is difficult to determine. The distribution appropriate to this part of the curve may depend quite critically on the nature of the system itself. Manufacturers will frequently subject their product to a burn-in period attempting to eliminate the early failures before lots are shipped to the consumer. Region II corresponds to "a constant failure rate function," and is the region of chance failures to which the exponential distribution applies. Region III corresponds to a wear-out process for which the normal distribution often provides an adequate model. For an actual system, the $\lambda(t)$-vs.-t curve may be quite different from that depicted in Figure X-8. For example, the exponential Region II may be entirely missing or the burn-in region may be negligible.

Returning to the failure rate function equations, it is convenient to solve (X-34) explicitly for both F(t) and f(t). This is accomplished by writing equation (X-34) in the form

$$\lambda(t)dt = -\frac{[-F'(t)\,dt]}{1-F(t)} \qquad (X\text{-}35)$$

where

$$F'(t) = \frac{d\,F(t)}{dt}.$$

Integrating both sides of equation (X-35) we have

$$-\int_0^t \lambda(x)\,dx = \ln[1-F(t)],$$

which is equivalent to

$$1 - F(t) = \exp\left[-\int_0^t \lambda(x)dx\right],$$

and therefore $F(t) = 1 - \exp\left[-\int_0^t \lambda(x)dx\right].$ \qquad (X-36)

If we differentiate equation (X-36), we obtain

$$f(t) = \lambda(t)\exp\left[-\int_0^t \lambda(x)dx\right]. \qquad (X\text{-}37)$$

The reader should convince himself that if we put $\lambda(t) = \lambda = $ constant in (X-36) and (X-37), we obtain

$$F(t) = 1 - e^{-\lambda t} \text{ and } f(t) = \lambda e^{-\lambda t},$$

which is just the exponential distribution. For the exponential distribution, then, the failure rate function is a constant (independent of t) and

$$\lambda(t)dt = P \text{ [failure occurs between t and t + dt | no prior failure]} = \lambda dt.$$

If we choose to describe a component failure distribution with the exponential distribution, we are assuming that we are on the constant, steady state portion of the bathtub curve with no burn-in or wear-out occurring. Because of the constancy of the failure rate function in this case, the exponential distribution is often referred to as the "random failure distribution," i.e., the probability of future failure is independent of previous successful operating time.

It is also valuable to note that if we use $e^{-t/\theta}$ to represent reliability, we are being conservative even if wear-out is occurring (but no burn-in), i.e., $R(t) \geqslant e^{-t/\theta}$ where $R(t)$ is the actual reliability and θ is the actual mean time to failure. This relation is true for $t \leqslant \theta$. (For a proof of this see Reference [15]).

Equations (X-36) and (X-37) may be used to investigate a wide variety of failure rate models. For instance, if $\lambda(t) = kt$ (linearly increasing failure rate) we find that

$$R(t) = 1 - F(t) = \exp(-kt^2/2)$$

which is known as the Rayleigh distribution. An important distribution of times-to-failure, the Weibull distribution, is obtained by putting $\lambda(t) = Kt^m$ $(m > -1)$ whence

$$f(t) = kt^m \exp\left(-\frac{kt^{m+1}}{m+1}\right)$$

and

$$R(t) = 1 - F(t) = \exp\left(-\frac{kt^{m+1}}{m+1}\right).$$

The Weibull distribution is a two-parameter distribution, where k is known as the scale parameter and m as the shape parameter. For m = 0 we have the exponential distribution and as m increases a wear-out behavior is modeled. When m increases to 2, $f(t)$ approaches normality. When m is less than 0 but greater than -1, the burn-in portion of certain bathtub curves may be modeled. Thus, by changing the value of m, we can use the Weibull distribution to embrace regions I, II, or III of the bathtub curve. The reader can find a more detailed description of the Weibull distribution in the literature of reliability (see reference [23] pp. 137-138, [32], Appendix D, and [36], p. 190 et seq.).

9. An Application Involving Time-To-Failure Distribution

The concept of a distribution of times-to-failure is an extremely important one and in order to impress this more thoroughly on the reader's mind, we consider the following example.

Similar components are being purchased from two manufacturers, A and B. Manufacturer A claims a mean life of 100 hours (θ_A = 100 hrs.) and states that the distribution of times-to-failure for his components is exponential. B also claims a mean life of 100 hours (θ_B = 100 hrs.) but states that the appropriate distribution of times-to-failure in his case is normal with a mean of 100 hours and a standard deviation of 40 hours.

For both types of components let us attempt to compute the reliability for 10 hours of operation. First we consider the component from manufacturer A.

$$R_A(t) = e^{-t/\theta_A}$$

$$R_A(10) = e^{-10/100} = e^{-0.1} = 0.905$$

Thus for Manufacturer A there is a 90.5% reliability.

Now we consider the components from manufacturer B. The distribution here is normal; we need to find a value of the transformed variate z corresponding to t = 10 hours.

$$z = \frac{t - \theta_B}{\sigma_B} = \frac{10 - 100}{40} = -2.25$$

This value of z cuts off a tail area of 0.01222 (from standard normal tables) and represents the probability of failure prior to t = 10 hours.* Thus

$$R_B(t = 10) = 1 - 0.01222 = 0.988$$

Thus for Manufacturer B the reliability is 98.8%.

From the above example we note the difference between R_A and R_B despite the fact that $\theta_A = \theta_B$. This difference arises because of the difference in the failure distributions. As t increases, eventually the exponential will give higher values of reliability than the normal. For instance, for t = 100 hours, R_A = 36.8% and R_B = 50.0% but for t = 200 hours, R_A = 13.5% and R_B = 0.62%. Somewhere between t = 100 hours and t = 200 hours both distributions give the same value for reliability. The reader may care to calculate the value of t that gives $R_A = R_B$.

10. Statistical Estimation

Suppose that we were engaged in a study of the heights of men between the ages of 20 and 30 in the greater Los Angeles area. This is a very large population, and although we might be eager and willing to measure every member of the population, this would assuredly prove to be physically impossible.

By way of compromise, we take a random sample from the population. The importance of the sample's being random will be discussed shortly. From the sample we can estimate any sample parameters that may be of interest, such as the sample mean, the sample median, the sample variance, and so forth. The problem is simply

*This is not exactly correct because values of z less than −2.5 correspond to negative values of time. However, the area in the tail corresponding to z = −2.5 is extremely small and we have not considered it necessary to make the small correction involved.

this: how good are these sample statistics as estimates of corresponding population quantities? As a matter of fact, have we any assurance that a sample mean is a better estimate of the population mean than is, for instance, a sample median or a sample mid-range? To answer questions such as these, we have to know precisely what is meant statistically by statements like:

"$\hat{\theta}_a$ is a good estimator of population parameter θ"

"$\hat{\theta}_b$ is a best estimator of population parameter θ"

"$\hat{\theta}_b$ is a better estimator of θ than is $\hat{\theta}_c$"*

These matters will be taken up in Section 13. First we must discuss the importance of choosing random samples, and then we must establish the concept of a sampling distribution, specifically the sampling distribution of the mean.

11. Random Samples

A random sample can be defined as a sample in which every member of the population has the same chance of being included. Most statistical calculations are based on the assumption of randomness; conclusions drawn from a sample which is thought to be random but which actually does not accurately reflect the characteristics of the population, can, therefore, be seriously in error.

A classic case in which the assumption of randomness was invalid involved the Literary Digest Poll of 1936. The poll attempted to sample public opinion on the subject of whether Mr. Roosevelt or Mr. Landon would win election to the presidency of the United States. The poll predicted a Landon victory whereas, in fact, Roosevelt won with a popular plurality of 11,069,785 votes and an electoral vote of 523 to 8. In this case, polling was carried out largely by means of telephone. In the depressed economic state of the country in those days, most of the telephones were owned by wealthy to moderately wealthy Republicans who intended to vote for Landon. The conclusions drawn from this non-random sample were spectacularly in error. Shortly thereafter the Literary Digest became defunct.

If you wish to select a random sample of electronic components from a large crate that has just been delivered, it would be incorrect to select only from those components at the top of the crate. In this case you might remain blissfully unaware that the crate had been dropped in transit and that every component at the botton of the crate had been smashed. Whenever sampling is performed, care must be taken to ensure a random sample because usual estimation techniques are based on this assumption. One simple method, for example, is to use random number tables to select the specific samples. Other random sample approaches are described in Reference [10].

12. Sampling Distributions

Suppose we take a random sample of size n from some population and compute a sample mean $\overline{x}_1$, where $\overline{x}_1 = \frac{1}{n} \sum_{i=1}^{n} x_i$. We could now take a second sample of size n

---

*The symbol $\hat{\theta}$ is added to designate an estimator of some population parameter θ.

and compute its mean $\bar{x}_2$. In a like manner we could generate other sample means, $\bar{x}_3$, $\bar{x}_3$, $\bar{x}_5$, etc. We do not expect these sample means to all be the same. In fact, these sample means are values of a random variable. Let the sample means in general be represented by the symbol $\bar{X}$ which denotes the random variable. The question arises, how is $\bar{X}$ distributed? A partial answer is provided by so-called <u>restricted central limit theorem</u> which states:

> If **X** (the random variable of interest) is distributed normally with mean μ and variance σ^2, then $\bar{X}$ is distributed normally with mean $\mu_{\bar{X}} = \mu$ and variance $(\sigma_{\bar{X}})^2 = \sigma^2/n$ where n is the sample size.

This theorem is exactly true only for populations of infinite size; for finite populations of size N and for samples of size n,

$$(\sigma_{\bar{X}})^2 = \frac{\sigma^2}{n}\left(\frac{N-n}{N-1}\right).$$

More important is the general central limit theorem which states that if X is distributed with mean μ and variance σ^2 but with distribution otherwise unknown, the distribution of $\bar{X}$ is still very closely approximated by the normal distribution with mean μ and variance σ^2/n, at least for large n (e.g., $n \geqslant 50$).

Consequently, whenever we deal with sample means of large samples, we are concerned with normal distributions. The variance of the sampling distribution of the mean decreases as the sample size increases and this provides a justification for taking as large a sample as possible. Note that the z transformation corresponding to an $\bar{x}$ value is

$$z = \frac{\bar{x} - \mu}{\sigma / \sqrt{n}}$$

Other estimators (e.g., median, range, variance, etc.) are characterized by their corresponding sampling distributions, most of which can be found in advanced texts on statistics, e.g., Reference [30]. For instance, for a normal distribution, the variance estimate, s^2, is a function of a value χ^2 from the chi square distribution through the relation

$$\chi^2 = \frac{(n-1)\,s^2}{\sigma^2}.$$

The chi square distribution has been extensively tabulated and is widely used in decision criterion, in goodness of fit tests, and in hypothesis testing.

13. Point Estimates—General

A single numerical value (such as $\bar{x}$) calculated from a sample constitutes a point estimate of some corresponding parameter. The associated random variable representing the collection of values is called the sample estimator. To facilitate the discussion let us use the symbol θ to designate the population parameter being estimated and the symbols $\widehat{\theta}_a$, $\widehat{\theta}_b$, $\widehat{\theta}_c$, ... to designate various sample estimators for θ. For instance, if θ were the population mean, then $\widehat{\theta}_a$ could be the sample mean estimator; $\widehat{\theta}_b$, the sample median estimator; $\widehat{\theta}_c$, the sample mid-range estimator, and so forth. The estimators $\widehat{\theta}_a$, $\widehat{\theta}_b$, $\widehat{\theta}_c$, ... will all have sampling distributions. The reader should note that the following characteristics of estimators relate to these sampling distributions.

a)   Unbiased Estimators

An estimator is said to be <u>unbiased</u> if its sampling distribution has a mean that equals the population parameter being estimated. Thus, if $\widehat{\theta}_a$ is an unbiased estimator of the population mean μ, then

$$E\left(\widehat{\theta}_a\right) = \mu.$$

From the properties of the expectation, we know that the sample mean $\overline{X}$ is an unbiased estimator of μ because $E(\overline{X}) = \mu$. On the other hand

$$S^2 = \frac{1}{n}\sum_{i=1}^{n}(X_i - \overline{X})^2$$

is a biased estimator of σ^2. If we multiply by $n/(n-1)$—Bessel's correction—we have

$$S^2 = \frac{1}{n-1}\sum_{i=1}^{n}(X_i - \overline{X})^2,$$

which is an unbiased estimator of σ^2.

b)   Minimum Mean Square Error Estimators
and Minimum Variance Estimators

The mean square error (MSE) of an estimator is defined as

$$MSE = E\left(\widehat{\theta} - \theta\right)^2. \qquad\qquad (X\text{-}38)$$

The MSE thus is a measure of the amount an estimator $\widehat{\theta}$ deviates from the true value θ. By adding and subtracting $E(\widehat{\theta})$ inside the parenthesis in equation (X-38), and manipulating the result, we can always rewrite the MSE as

$$MSE = E\left[\widehat{\theta} - E(\widehat{\theta})\right]^2 + \left[E(\widehat{\theta}) - \theta\right]^2.$$

The first term on the right-hand side is the variance of the estimator and the second term is called the square of the bias of the estimator. If the estimator is unbiased then $E(\widehat{\theta}) = \theta$ and

$$\text{MSE} = \text{E}\left[\hat{\theta} - \text{E}(\hat{\theta})\right]^2.$$

Hence the MSE for an unbiased estimator is simply the variance of the estimator.

If among several estimators $\hat{\theta}_a, \hat{\theta}_b, \hat{\theta}_c, \ldots$, one of them has the smallest MSE, then that estimator is called the minimum mean square error (MMSE) estimator. If among several estimators, all of which are unbiased, one has minimum variance, then that estimator is called the minimum variance unbiased estimator and abbreviated MVUE.

The choice of estimator depends on the situation. If we are going to use an estimator in numerous applications then we generally want the estimator to be unbiased, because on the average we want the estimator to equal the true value. If we choose between two or more unbiased estimators, we usually choose the one with minimum variance. In comparing the variance of two unbiased estimators $\hat{\theta}_1$ and $\hat{\theta}_2$, we say that the one with the smaller variance is relatively more efficient and in fact, we use the ratio

$$\frac{\text{var}\left(\hat{\theta}_2\right)}{\text{var}\left(\hat{\theta}_1\right)}$$

as one measure of the relative efficiency of estimator $\hat{\theta}_1$ with respect to $\hat{\theta}_2$.

If, however, we are going to use an estimator only once or a few times, then a (biased) MMSE estimator may be more efficient. In this case we are more interested in minimizing the deviation from the true value than we are in the long-run unbiased property.

c) **Consistent Estimators**

If $\hat{\theta}_a$ is a consistent estimator of θ, then

$$P\left[|\hat{\theta}_a - \theta| < \epsilon\right] > (1-\delta) \text{ for all } n > n'$$

where ϵ and δ are arbitrarily small positive numbers and n' is some integer. We can interpret the above equation as saying that as the sample size n increases, the pdf of the estimator concentrates about the true value of the parameter. When n becomes very large then the probability that the estimator deviates from the true value goes to zero. In this case we say that "$\hat{\theta}_a$ converges in probability to θ."

Properties (a), (b), and (c) are the principal characteristics that determine whether an estimator is good or not. Further discussion of estimator considerations is given in reference [24].

14. Point Estimators—Maximum Likelihood

There is one very important technique for calculating estimators that is called the method of maximum likelihood. This method is widely used, for example, in computing parameter estimates in life testing. For large sample sizes (n→∞) under rather general conditions, the maximum likelihood technique yields estimators which are consistent and which are both MMSE and MVUE. Even for moderate or small samples, the maximum likelihood technique yields estimates which are generally usable. The technique is based on the supposition that the particular random sample drawn from the population is the most likely one that could have been chosen. To make this supposition reasonable, consider two examples.

Bridge players do not expect to pick up a hand containing all 13 spades. The probability of such a hand is extremely small because it can arise in only one way. The probability of a 13-spade hand, however, is the same as that associated with any hand that is preassigned card by card, because that hand can also arise in only one way. The kind of hand you usually get may look something like this:

4 Spades
2 Hearts
4 Diamonds
3 Clubs

This hand can arise in an enormous number of ways—to be precise, $\binom{13}{4}\binom{13}{2}\binom{13}{4}\binom{13}{3} = (13)^4 (11)^3 (10)^2 (3)$ ways (over 10 billion). In fact, you will get the distribution 4-4-3-2 (regardless of suit) about 20% of the time. For most of the rest of the time you will be getting hands that are very similar to this. The maximum likelihood technique is based on the premise that the sample we have is a most probable one, or is near the most probable one.

As a more physical illustration, assume that we had a special camera that photographed the molecules in a box full of gas. When the photographs are developed, not only do we see the actual, instantaneous locations of the molecules but also their individual vector velocities. We could take photographs like this for thousands of years and they would all look pretty much the same: a homogeneous distribution of molecules in space going all directions. Even so, there is a positive (although miniscule) probability that we shall find a picture showing all the particles crowded into one corner of the box and all heading due north. If we were to apply the maximum likelihood technique to one sample (a given photograph), then we would be making the assumption that the sample was a probable one and not the very unlikely one.

In general, the technique of maximum likelihood is founded on the assumption that our sample is representative of the most likely one we could have withdrawn from the population—always with the proviso that we have gone to some pains to assure that it is a random one.

Suppose that we are taking random samples from a population that is distributed according to the pdf $f(x; \theta)$ where θ is some unknown population parameter that we desire to estimate. Assume that our sample (size n) is $x_1, x_2, \ldots, x_n$ and that the sample variables are independent. Using the pdf, we write down an expression that gives us the probability associated with this particular sample and then apply the condition that it be a maximum.

The probability that out first reading will be x_1 in the interval dx_1 is simply $f(x_1;\theta)dx_1$. The probability that our first reading will be x_1 in dx_1 <u>and</u> that our second reading will be x_2 in dx_2 is

$$f(x_1;\theta)dx_1 \cdot f(x_2;\theta)dx_2$$

Following this line of reasoning, we can now write down an expression that gives us the probability of our sample:

$$P[\text{sample}] = f(x_1;\theta)dx_1\, f(x_2;\theta)dx_2\, f(x_3;\theta)dx_3 \ldots f(x_n;\theta)dx_n \quad (X\text{-}39)$$

If we drop off the differentials we get an expression that is known as the <u>likelihood function</u>:

$$\text{Likelihood Function} = f(x_1;\theta)\, f(x_2;\theta) \ldots f(x_n;\theta) = \prod_{i=1}^{n} f(x_i;\theta) \quad (X\text{-}40)$$

where the symbol Π stands for a continued product. The likelihood function is no longer equal to the probability of the sample but it represents a quantity that is proportional to that probability.*

The next step is to take the natural logarithm ("ln") of the likelihood function. This is simply a matter of convenience because most of the pdf's we encounter are exponential in form and it is easier to work with the natural logarithm than with the function itself.

$$\ln(\text{Likelihood Function}) = L(\theta) = \ln \prod_{i=1}^{n} f(x_1;\theta) = \sum_{i=1}^{n} \ln f(x_i;\theta) \quad (X\text{-}41)$$

Notice that this is a function only of θ because all the x_i's (our sample values) are known. We are now in a position to determine θ for which $L(\theta)$ is a maximum. We do this by setting the derivative of $L(\theta)$ with respect to θ equal to zero and solving for θ.

$$\frac{d}{d\theta} L(\theta) = 0. \quad (X\text{-}42)$$

Assuming a solution is obtainable (which it generally is), the result is written θ_{ML}, the maximum likelihood estimate of the unknown population parameter θ.

We shall now consider some specific examples to indicate how the maximum likelihood technique is applied in practice. Suppose that we are taking random samples from a population whose distribution is normal with unknown mean μ and known variance $\sigma^2 = 1$.

---

*If the sample variables were not independent, then the likelihood would consist of a multivariate distribution, and we would attempt to maximize that (see Reference [24]).

$$f(x; \mu, \sigma = 1) = \frac{1}{\sqrt{2\pi}} \exp\left[-\frac{(x-\mu)^2}{2}\right].$$

We desire to make a maximum likelihood estimate of μ.

The likelihood function is

$$\frac{1}{\sqrt{2\pi}} \exp\left[-\frac{(x_1-\mu)^2}{2}\right] \frac{1}{\sqrt{2\pi}} \exp\left[-\frac{(x_2-\mu)^2}{2}\right] \cdots \frac{1}{\sqrt{2\pi}} \exp\left[-\frac{(x_n-\mu)^2}{2}\right]$$

$$= \frac{1}{(2\pi)^{n/2}} \exp\left[-\frac{1}{2}\sum_{i=1}^{n}(x_i-\mu)^2\right].$$

Taking natural logarithms we have

$$L(\mu) = -\frac{n}{2}\ln(2\pi) - \frac{1}{2}\sum_{i=1}^{n}(x_i-\mu)^2.$$

Applying the maximization condition,

$$\frac{dL(\mu)}{d\mu} = \left(-\frac{1}{2}\right)(2)\sum_{i=1}^{n}(x_i-\mu)(-1) = \sum_{i=1}^{n}(x_i-\mu) = 0.$$

This yields

$$\Sigma x_i - n\mu = 0,$$

so

$$\mu_{ML} = \frac{1}{n}\sum_{i=1}^{n}x_i = \overline{x}.$$

Thus, the maximum likelihood estimate of μ is just the ordinary arithmetic mean.

If more than one population parameter is to be estimated, the process is similar. Suppose that in our previous example we knew neither μ nor σ^2. Then our fundamental pdf is

$$f(x; \mu, \sigma^2) = \frac{1}{\sqrt{2\pi}\,\sigma} \exp\left[-\frac{(x-\mu)^2}{2\sigma^2}\right].$$

The likelihood function becomes

$$f(x_1, \mu, \sigma^2) \cdots f(x_n; \mu, \sigma^2) = \frac{1}{(2\pi\sigma^2)^{n/2}} \exp\left[-\sum_{i=1}^{n}\frac{(x_i-\mu)^2}{2\sigma^2}\right].$$

Taking natural logarithms we have

$$L(\mu, \sigma^2) = -\frac{n}{2} \ln (2\pi) - \frac{n}{2} \ln \sigma^2 - \frac{1}{2\sigma^2} \sum_{i=1}^{n} (x_i - \mu)^2 .$$

We now find $\partial L/\partial \mu$ and $\partial L/[\partial(\sigma^2)]$ and set the results equal to zero. The first operation yields the same result as before, namely,

$$\mu_{ML} = \frac{1}{n} \sum_{i=1}^{n} x_i = \bar{x}.$$

The second operation yields

$$\sigma^2_{ML} = \frac{1}{n} \sum_{i=1}^{n} (x_i - \bar{x})^2 .$$

This is a <u>biased</u> estimate of σ^2. The bias may be removed by multiplying σ^2_{ML} by the quantity $n/(n-1)$. Then

$$\sigma^2_{unbiased} = \frac{1}{n-1} \sum_{i=1}^{n} (x_i - \bar{x})^2 .$$

There is little difference between σ^2_{ML} and $\sigma^2_{unbiased}$ if n, the sample size, is large (say $n \geqslant 30$). The estimated value of the variance is often denoted by the symbol s^2.

As a final example let us return to the exponential distribution and find the ML estimate for θ, the mean life.

$$f(t; \theta) = \theta^{-1} e^{-t/\theta} \quad (t > 0).$$

Assume we have observed n times of failure for n components tested. Then

$$f(t_1; \theta) \ldots f(t_n; \theta), = \theta^{-n} \exp\left[-\frac{1}{\theta} \sum_{i=1}^{n} t_i \right]$$

and

$$L(\theta) = - n \ln \theta - \frac{1}{\theta} \sum_{i=1}^{n} t_i .$$

Therefore,

$$\frac{dL}{d\theta} = -\frac{n}{\theta} + \frac{1}{\theta^2} \sum_{i=1}^{n} t_i = 0$$

and

$$\frac{1}{\theta} \sum_{i=1}^{n} t_i = n$$

so that

$$\theta_{ML} = \frac{1}{n} \sum_{i=1}^{n} t_i, \text{ again the simple arithmetic mean.}$$

15. Interval Estimators

We have seen in the previous sections how we can make point estimates of population parameters on the basis of one or more random samples drawn from the population. We can, if we desire, adopt a different approach. This involves making an assertion such as:

$$P\left[\left(\widehat{\theta}_{lower} < \theta < \widehat{\theta}_{upper}\right)\right] = \eta$$

where θ is some unknown population parameter, $\widehat{\theta}_{lower}$ and $\widehat{\theta}_{upper}$ are estimators associated with a random sample and η is a probability value such as 0.99, 0.95, 0.90, etc. If, for instance, $\eta = 0.95$ we refer to the interval

$$\left(\theta_L < \theta < \theta_U\right)$$

for particular values of $\widehat{\theta}_{lower}$ and $\widehat{\theta}_{upper}$ as a 95% confidence interval. In this case we are willing to accept a 5% probability (risk) that our assertion is not, in fact, true.

To help clarify the concept of a confidence interval we can look at the situation in a geometrical way. Suppose we draw repeated samples (x_1, x_2) from a population, one of whose parameters, θ, we desire to bracket with a confidence interval. We construct a three-dimensional space with the vertical axis corresponding to θ and with the two horizontal axes corresponding to values of x_1 and x_2 (see Figure X-9). The actual value of the population parameter θ is marked on the vertical axis and a horizontal plane is passed through this point. Now we take a random sample (x_1, x_2) from which we calculate the values θ_U and θ_L at, say, the 95% confidence level. The interval defined by θ_U and θ_L is plotted on the figure.

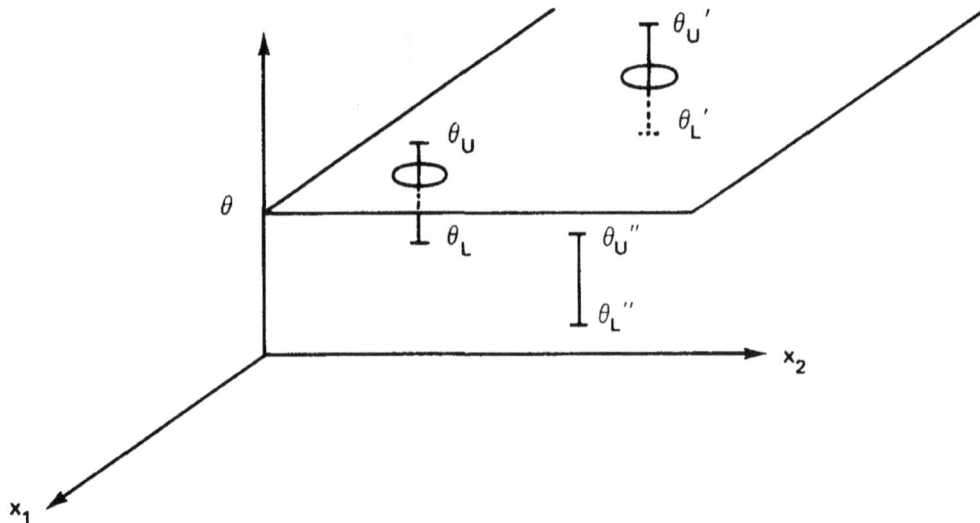

Figure X-9. Geometrical Interpretation of the Concept
of a Confidence Interval

Next we take a second sample (x_1', x_2') from which we calculate the values θ_U' and θ_L' at the 95% level. This interval is plotted on the figure. A third sample (x_1'', x_2'') yields the values θ_U'' and θ_L'', etc. In this way we can generate a large family of confidence intervals. The confidence intervals depend only on the sample values (x_1, x_2), (x_1', x_2'), etc., and hence we can calculate these intervals without knowledge of the true value of θ. If the confidence intervals are all calculated on the basis of 95% confidence and if we have a very large family of these intervals, then 95% of them will cut the horizontal plane through θ (and thus include θ) and 5% of them will not.

The process of taking a random sample and computing from it a confidence interval is equivalent to the process of reaching into a bag containing thousands of confidence intervals and grabbing one at random. If they are all 95% intervals, our chance of choosing one that does indeed include θ will be 95%. In contrast, 5% of the time we will be unlucky and select one that does not include θ (like the interval (θ_U'', θ_L'') in Figure X-9). If a risk of 5% is judged too high, we can go to 99% intervals, for which the risk is only 1%. As we go to higher confidence levels (and lower risks) the lengths of the intervals increase until for 100% confidence the interval includes every conceivable value of θ (I am 100% confident that the number of defective items in a population of 10,000 is somewhere between 0 and 10,000.). For this reason 100% confidence intervals are of little interest.

Now we shall give an example showing how the quantities θ_L and θ_U are computed from a sample drawn from a normally distributed population with mean μ and standard deviation σ. For the example, we will assume that we wish to bracket μ and that we know σ (based on previous data and knowledge). If each sample is drawn from a normal distribution then the sample mean $\overline{X}$ has a normal distribution with mean μ and standard deviation $\sigma/\sqrt{n}$, where n is the sample size. Even if each sample value is not drawn from a normal distribution, then by the Central Limit Theorem, for sufficiently large n, $\overline{X}$ will still be approximately normal with mean μ and standard deviation $\sigma/\sqrt{n}$. The quantity Z will then be the standardized normal random variable, where

$$Z = \frac{\overline{X} - \mu}{\sigma/\sqrt{n}}$$

and where the distribution of Z is tabulated. Because the distribution of Z is tabulated we can find, for any assigned probability η, values $-w$ and w such that

$$P[-w < Z \leqslant w] = \eta.$$

For example, for $\eta = 0.95$, w = 1.96. Substituting for Z in the above equation we thus have

$$P\left[-w \leqslant \frac{\overline{X} - \mu}{\sigma/\sqrt{n}} \leqslant w\right] = \eta$$

where again w is known for any given η. (For example, we could substitute the values 0.95 for η and 1.96 for w.)

We are now going to concentrate on the inequality on the left-hand side of the last equation and attempt to convert it into the form

$$\left[\theta_L < \mu < \theta_U\right].$$

Let us first multiply through by the factor $\sigma/\sqrt{n}$ thus converting the inequality to

$$\left[-w\frac{\sigma}{\sqrt{n}} < \overline{X}-\mu < +w\frac{\sigma}{\sqrt{n}}\right].$$

Now we subtract $\overline{X}$ from every term of the inequality

$$\left[-w\frac{\sigma}{\sqrt{n}} - \overline{X} < -\mu < w\frac{\sigma}{\sqrt{n}} - \overline{X}\right].$$

Next we multiply through by -1 remembering, in the process, to reverse the directions of the inequalities.

$$\left[w\frac{\sigma}{\sqrt{n}} + \overline{X} > \mu > -w\frac{\sigma}{\sqrt{n}} + \overline{X}\right]$$

which we can write in the form

$$\left[\overline{X} - w\frac{\sigma}{\sqrt{n}} < \mu < \overline{X} + w\frac{\sigma}{\sqrt{n}}\right].$$

Substituting a particular value $\overline{x}$ for the random variable $\overline{X}$ we thus have a particular interval,

$$\overline{x} - w\frac{\sigma}{\sqrt{n}} < \mu < \overline{x} + w\frac{\sigma}{\sqrt{n}}.$$

The above inequality yields our desired confidence interval for μ. In the case of the population mean, then

$$\theta_L = \overline{x} - w\frac{\sigma}{\sqrt{n}} \text{ and } \theta_U = \overline{x} + w\frac{\sigma}{\sqrt{n}}.$$

If a confidence coefficient η has been assigned, $\overline{x}$, n, and w are known quantities. The value of σ is also assumed to be known. If we do not know the value of σ then we can, as described earlier for the normal distribution, estimate σ from our sample thus obtaining the quantity s. Now we can form the standardized value t, where

$$t = \frac{\overline{x} - \mu}{s/\sqrt{n}}.$$

Proceeding as we did with the **Z** variable, we can form the inequality

$$\overline{x} - \frac{ts}{\sqrt{n}} < \mu < \overline{x} + \frac{ts}{\sqrt{n}}.$$

The value t, however, is not a value from the standard normal distribution. The distribution was investigated in the early part of this century by W.S. Gossett and is known as the t distribution; its properties are now extensively tabulated. So if σ is unknown, we find the value of t for given η from t tables and not from normal tables. The values of the t distribution depend on the sample size (degrees of freedom). As it turns out, when the sample size is greater than 25 or 30 the t table value is indistinguishable from the normal table value so that in that case, normal tables may be used.

In terms of estimating the reliability or mean time to failure, one-sided confidence intervals are more commonly seen than two-sided intervals. If the sampling distribution is symmetric (equal tail areas on the ends), then a two-sided interval may easily be converted into a corresponding one-sided interval; e.g., if

$$0.95 < R < 0.98$$

at the 95% confidence level, then

$$R > 0.95$$

at the 97.5% confidence level.

For the exponential distribution, the mean time between failures, θ', was estimated by the point estimate

$$\theta_{ML} = \frac{1}{n} \sum_{i=1}^{n} t_i$$

where t_i are the observed times of failure. It can be shown that $\chi^2 = \dfrac{n\theta_{ML}}{\theta}$ follows a chi square distribution with 2n degrees of freedom. Letting χ^2 (97.5, 2n) and χ^2 (2.5, 2n) be the chi square values corresponding to cumulative distribution values of 97.5% and 2.5%, we have for a two-sided 95% confidence interval

$$\chi^2(2.5, 2n) < \frac{n\theta_{ML}}{\theta} < \chi^2(97.5, 2n)$$

or, equivalently,

$$\frac{n\theta_{ML}}{\chi^2(97.5, 2n)} < \theta < \frac{n\theta_{ML}}{\chi^2(2.5, 2n)}.$$

Other intervals for different levels of confidence can be obtained from the tabulated values of the chi square distribution. Confidence intervals for the failure rate $\lambda = 1/\theta$ can be easily obtained by inverting the above inequalities for θ.

For more involved experiments, the confidence intervals, as well as the maximum likelihood point estimates, depend on the way the data are collected. In what is called a type 1 test, for example, n components having the same failure rate are operated for a prearranged time interval T. The number of components failing in this time interval is then random. For a type 2 test, n components are operated until a preassigned number of components fail and this preassigned number may be less than n.

The text by Mann, Schafer, and Singpurwalla (see reference [24]), cited earlier, further discusses tests of type 1 and 2, giving associated point estimates and confidence intervals. Testing with replacement is also discussed and confidence intervals and point estimates are given for the Weibull and gamma distributions as well as the exponential distribution under a variety of circumstances.

16. Bayesian Analyses

In the past discussions, we have been treating the parameters of the sampling distributions as being fixed. In many applications, this is a questionable assumption. In the Bayesian approach, the parameters of the sampling distribution are treated not as fixed, but as random variables. With regard to the exponential distribution $f(x) = \frac{1}{\theta} e^{-x/\theta}$, the mean time to failure θ is thus treated as itself being associated with a probability distribution. Expressing the exponential in terms of the failure rate $\lambda = 1/\theta$, we have $f(x) = \lambda e^{-\lambda x}$. The failure rate is thus also associated with a probability distribution (because of the relation $\lambda = 1/\theta$, the distribution of λ is given by the distribution for θ and vice versa). From here on λ and θ will denote the random variables associated with the parameters λ and θ.

Let the pdf for λ be denoted by $p(\lambda)$. The pdf $p(\lambda)$ is known as the prior distribution and it represents our prior knowledge of λ before a given sample is taken. Assume now that a given sample $(t_1, t_2, \ldots, t_n)$ of component failure times is collected. We then talk about the posterior distribution of λ which represents our updated knowledge of the distribution of λ with the additional sample data incorporated.

The posterior distribution of λ, whose pdf we shall denote by $p(\lambda|D)$, is easily obtained by applying Bayes theorem* (the symbol "D" denotes the data sample, e.g., $(t_1, t_2, \ldots, t_n)$). Now Bayes theorem says

$$P(B|A) = \frac{P(A/B)\, P(B)}{\sum_{B} P(A/B)\, P(B)}.$$

Letting "A" denote the data sample "D," and "B" the event that the failure rate lies between λ and $\lambda + d\lambda$, we have

$$p(\lambda|D) = \frac{\exp\left[-\sum_{i=1}^{n} \lambda t_i\right] \lambda^n\, p(\lambda)}{\exp\left[-\sum_{i=1}^{n} \lambda t_i\right] \lambda^n\, p(\lambda)\, d\lambda}.$$

---

*See Chapter VI, Section 9.

Here we have replaced the summation sign by the integral sign. Because the denominator does not involve λ (it is integrated out) we can write the above equation as

$$p(\lambda|D) = K \exp\left[-\sum_{i=1}^{n} \lambda t_i\right] \lambda^n p(\lambda)$$

where K is treated as a normalizing constant. The distribution $p(\lambda|D)$ is the posterior distribution of λ now incorporating now both our prior knowledge and the observed data sample.

Bayes theorem thus gives a formal way of updating information about the failure rate λ (i.e., going from $p(\lambda)$ to $p(\lambda|D)$). If a second sample D' were collected (say t_i', $t_2', \ldots, t_n'$) then the distribution of λ could be updated to incorporate both sets of data. If $p(\lambda|D,D')$ represents the posterior distribution of λ based on both sets of data D and D' then we simply use the above equation with $p(\lambda|D)$ now as our prior giving

$$p(\lambda|D, D') = K \exp\left[-\sum_{i=1}^{n} \lambda t_i\right] \lambda^n p(\lambda|D)$$

Choices for the initial prior $p(\lambda)$ as well as techniques for handling various kinds of data are described in detail in various texts (see references [24] and [30]). In Bayesian approaches, the pdf's obtained for the parameters (such as $p(\lambda|D)$) give detailed information about the possible variability and uncertainty in these parameters. We can obtain point values, such as the most likely value of λ or the mean value of λ. We can also obtain interval values, which are probability intervals and are sometimes called Bayesian confidence intervals. For example, having determined $p(\lambda|D)$ we can then determine lower and upper 95% values of λ_L and λ_U, such that there is a 95% probability that the failure rate lies between these values, i.e.,

$$\int_{\lambda_L}^{\lambda_U} p(\lambda|D)\, d\lambda = 0.95.$$

Other bounds and other point values can be obtained in the Bayesian approach because the parameter distribution (e.g., $p(\lambda|D)$) is entirely known and this distribution represents our knowledge about the parameter. The Bayesian approach has the advantage that engineering experience and general knowledge, as well as "pure" statistical data, can be factored into the prior distribution (and hence posterior distribution). Once distributions are obtained for each relevant component parameter, such as the component failure rate, then the distribution is straightforwardly obtained for any system parameter quantified in the fault tree, such as the system unavailability, reliability, or mean time to failure. One must be very careful in determining the priors which truly represent the analyst's knowledge and in ascertaining the impact of different priors if they are all potentially applicable. Bayesian approaches are further discussed in reference [24].

CHAPTER XI – FAULT TREE EVALUATION TECHNIQUES

1. Introduction

This chapter describes the techniques which form the bases for manual and automated fault tree evaluations and discusses basic results obtained from these evaluations. Once a fault tree is constructed it can be evaluated to obtain qualitative and/or quantitative results. For simpler trees the evaluations can be performed manually; for complex trees computer codes will be required. Chapter XII discusses computer codes which are available for fault tree evaluations.

Two types of results are obtainable in a fault tree evaluation: qualitative results and quantitative results. Qualitative results include: (a) the minimal cut sets of the fault tree, (b) qualitative component importances, and (c) minimal cut sets potentially susceptible to common cause (common mode) failures. As previously discussed, the minimal cut sets give all the unique combinations of component failures that cause system failure. The qualitative importances give a "qualitative ranking" on each component with regard to its contribution to system failure. The common cause/common mode evaluations identify those minimal cut sets consisting of multiple components which, because of a common susceptibility, can all potentially fail due to a single failure cause.

The quantitative results obtained from the evaluation include: (a) absolute probabilities, (b) quantitative importances of components and minimal cut sets, and (c) sensitivity and relative probability evaluations. The quantitative importances give the percentage of time that system failure is caused by a particular minimal cut set or a particular component failure. The sensitivity and relative probability evaluations determine the effects of changing maintenance and checking times, implementing design modifications, and changing component reliabilities. Also included in the sensitivity evaluations are error analyses to determine the effects of uncertainties in failure rate data.

Listed below is a summary of the type of results obtained from a fault tree evaluation. In the following sections we discuss fault tree evaluation in more detail.

Classes of Results Obtained From a Fault Tree Evaluation:

Qualitative Results

a.	Minimal cut sets:	Combinations of component failures causing system failure
b.	Qualitative importances:	Qualitative rankings of contributions to system failure
c.	Common cause potentials:	Minimal cut sets potentially susceptible to a single failure cause

Quantitative Results

a.	Numerical probabilities:	Probabilities of system and cut set failures
b.	Quantitative importances:	Quantitative rankings of contributions to system failure

c.   Sensitivity evaluations: Effects of changes in models and data, error
                                        determinations

2.   Qualitative Evaluations

For the qualitative evaluations, the minimal cut sets are obtained by Boolean
reduction of the fault tree as previously described in Chapter VII, section 4.
Additional examples of minimal cut set determinations are given in this section to
further familiarize the reader with Boolean reduction techniques. The minimal cut
sets obtained are used not only in the subsequent qualitative evaluations but in all
the quantitative evaluations as well.

(a)   Minimal Cut Set Determinations

Because the minimal cut sets form the bases for all types of evaluations considered
here, we shall briefly review the determination of minimal cut sets from the tree. In
general, as stated in Section 4 of Chapter VII, our goal is to obtain the top event in
the minimal cut set form

$$T = M_1 + M_2 + M_3 + \ldots, + M_n.$$

The minimal cut sets M_j consists of combinations of primary failures (component
failures), e.g., $M_j = C_1 C_2 C_3$, which are the smallest combinations of primary failures
that cause system failure. The substitution can be either a top-down substitution or a
bottom-up substitution as previously described. Most computer algorithms for
determining minimal cut sets are based on these principles. (Computer codes are
discussed in Chapter XII.)

Let us now consider the pressure tank fault tree of Figure XI-1. Figure XI-1 is
similar to the detailed pressure tank fault tree constructed in Chapter VIII and shown
in Figure VIII-13. In Chapter VIII, minimal cut sets were determined for a reduced
version of the tree. We will determine here the minimal cut sets of the detailed tree as
a first type of qualitative and/or quantitative evaluation.

In the fault tree in Figure XI-1 we designate primary failures by $P_1, P_2, \ldots$ in
circles; secondary failures by $S_1, S_2 \ldots$ in diamonds; undeveloped events by E_1, E_2
$\ldots$ in diamonds; and all higher fault events by $G_1, G_2 \ldots$ except for the top event
which we designate T.

The set of Boolean equations equivalent to the fault tree of Figure XI-1 is:

$$T = P_1 + S_1 + G_1$$
$$G_1 = E_1 + G_2$$
$$G_2 = P_2 + S_2 + G_3$$
$$G_3 = G_4 \cdot G_5$$
$$G_4 = G_6 + G_7$$
$$G_5 = P_3 + S_3 + E_3$$
$$G_6 = P_4 + S_4 + E_4$$
$$G_7 = P_5 + S_5 + G_8$$
$$G_8 = P_6 + S_6 + E_6.$$

Note that we have one equation for each gate in the tree.

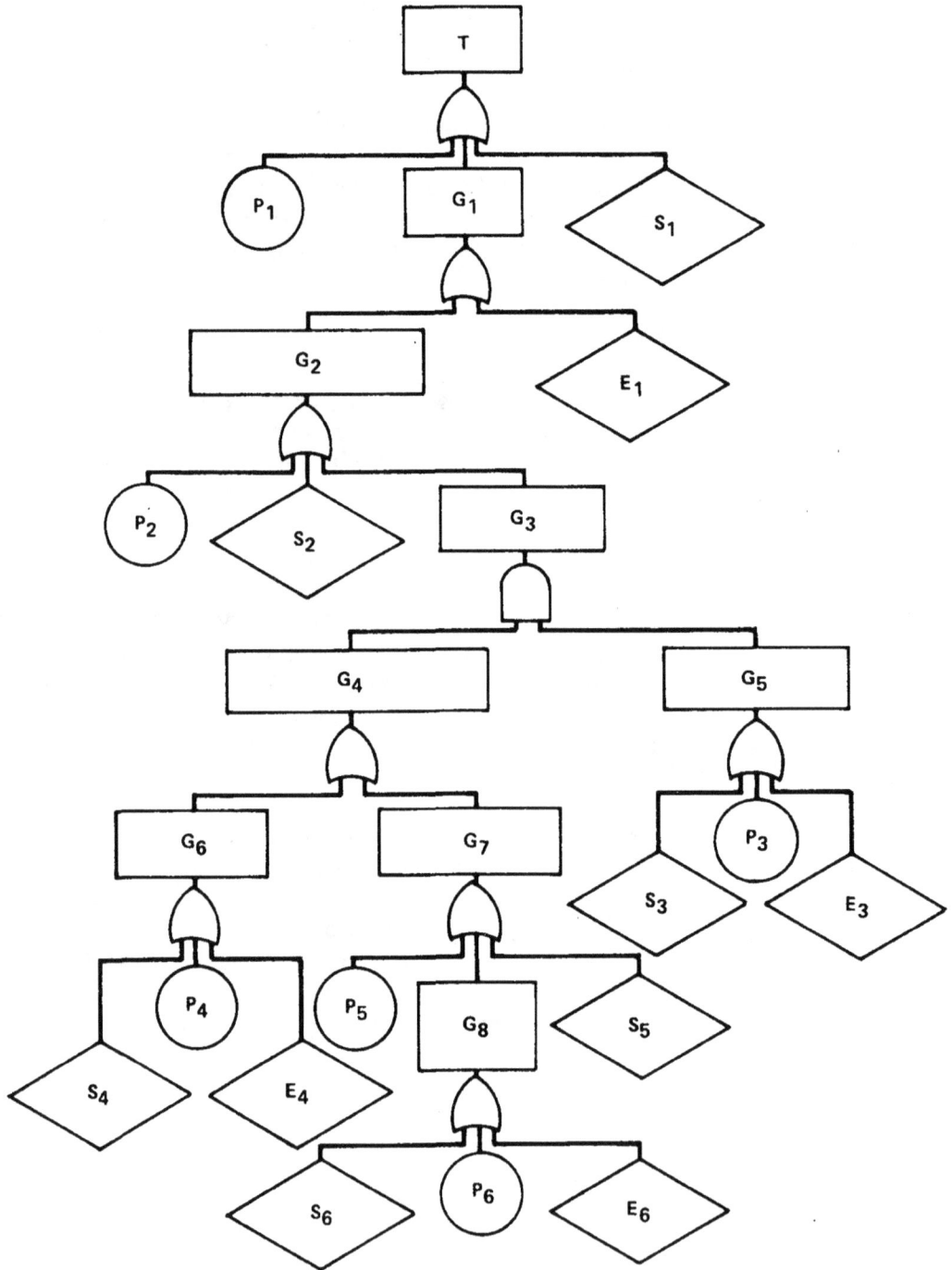

Figure IX-1. Pressure Tank Fault Tree

We will use the bottom-up procedure and write each gate equation in terms of primary events (the P's, E's, and S's) by substituting for the G's. Using the distributive law and the law of absorption we will then transform each gate equation to minimal cut set form. G_8 is already in minimal cut set form. G_7 contains only the higher fault G_8 and so substituting it becomes

$$G_7 = P_5 + S_5 + P_6 + S_6 + E_6$$

which is now in minimal cut set form. Both G_6 and G_5 are already in minimal cut set form. G_4 contains G_6 and G_7, which are already in minimal cut set form, and thus becomes

$$G_4 = P_4 + S_4 + E_4 + P_5 + S_5 + P_5 + S_6 + E_6.$$

Substituting for G_4 and G_5 in the equation for G_3 we have

$$G_3 = (P_4 + S_4 + E_4 + P_5 + S_5 + P_6 + S_6 + E_6) \cdot (P_3 + S_3 + E_3).$$

Using the distributive law we now express G_3 in its expanded form:

$$\begin{aligned}
G_3 = &(P_4 \cdot P_3) + (P_4 \cdot S_3) + (P_4 \cdot E_3) \\
&+ (S_4 \cdot P_3) + (S_4 \cdot S_3) + (S_4 \cdot E_3) \\
&+ (E_4 \cdot P_3) + (E_4 \cdot S_3) + (E_4 \cdot E_3) \\
&+ (P_5 . P_3) + (P_5 \cdot S_3) + (P_5 \cdot E_3) \\
&+ (S_5 \cdot P_3) + (S_5 \cdot S_3) + (S_5 \cdot E_3) \\
&+ (P_6 \cdot P_3) + (P_6 \cdot S_3) + (P_6 \cdot E_3) \\
&+ (S_6 \cdot P_3) + (S_6 \cdot S_3) + (S_6 \cdot E_3) \\
&+ (E_6 \cdot P_3) + (E_6 \cdot S_3) + (E_6 \cdot E_3).
\end{aligned}$$

Now $G_2 = P_2 + S_2 + G_3$ and consequently
$$G_1 = E_1 + P_2 + S_2 + G_3$$
so that $T = P_1 + S_1 + E_1 + P_2 + S_2 + G_3.$

Substituting the expanded form of G_3 into this last equation, we finally have T in correct minimal cut set form.

$$\begin{aligned}
T = &P_1 + S_1 + E_1 + P_2 + S_2 \\
&+ (P_4 \cdot P_3) + (P_4 \cdot S_3) + (P_4 : E_3) \\
&+ (S_4 \cdot P_3) + (S_4 \cdot S_3) + (S_4 \cdot E_3) \\
&+ (E_4 \cdot P_3) + (E_4 \cdot S_3) + (E_4 \cdot E_3) \\
&+ (P_5 \cdot P_3) + (P_5 \cdot S_3) + (P_5 \cdot E_3) \\
&+ (S_5 \cdot P_3) + (S_5 \cdot S_3) + (S_5 \cdot E_3) \\
&+ (P_6 \cdot P_3) + (P_6 \cdot S_3) + (P_6 \cdot E_3) \\
&+ (S_6 \cdot P_3) + (S_6 \cdot S_3) + (S_6 \cdot E_3) \\
&+ (E_6 \cdot P_3) + (E_6 \cdot S_3) + (E_6 \cdot E_3).
\end{aligned}$$

The top event, or system model, thus contains:
    5 single component minimum cut sets, and
    24 double component minimum cut sets.

We note that if the S's denote secondary failures, then we do not really need to explicitly show them. All we need to do is to redefine the primary failures P to denote <u>any</u> type of cause and we can, if we wish, separate the causes in the quantitative analyses. By deleting the S's in the fault tree, we delete all the cut sets containing S. We now have

       3 single component minimal cut sets, and

      10 double component minimal cut sets.

With larger trees, the savings in number of minimal cut sets will be even greater if we do not explicitly show the secondary failures.

(b) Qualitative Importances

After obtaining the minimal cut sets, some idea of failure importances can be obtained by ordering the minimal cut sets according to their size. The single-component minimal cut sets (if any) are listed first, then the double-component minimal cut sets, then the triple, etc. Computer codes usually list the minimal cut sets in this order.

Because required computer times can increase dramatically as the size of the minimal cut sets increase, it is often the practice to obtain only the single-, double-, and perhaps triple-component minimal cut sets. As an additional calculation, higher order minimal cut sets (quadruples, etc.) can also be obtained on a selective basis if they show potential susceptibility to common cause failures (discussed further in the next section).

Because the failure probabilities associated with the minimal cut sets often decrease by orders of magnitude as the size of the cut set increases, the ranking according to size gives a gross indication of the importance of the minimal cut set. For example, if individual component failure probabilities are of the order of 10^{-3}, a single-component cut set probability will be of the order of 10^{-3}, and a double cut set 10^{-6}, a triple 10^{-9}, etc. Component failure probabilities are in general different and depend on testing intervals, downtimes, etc.; therefore the ranking of minimal cut sets according to size gives only a general indication of importance.

The minimal cut set information can sometimes be used directly to check design criteria. For example, if a design criterion states that no single component shall fail the system, then this is equivalent to stating that the system shall contain no single component minimal cut sets. The minimal cut sets can be checked to see if this criterion is satisfied. Similar checks can be done on criteria that restrict specific types of failures which may not individually fail a system (e.g., active components or human errors).

(c) Common Cause Susceptibilities

The primary failures (component failures) on a fault tree do not necessarily have to be independent. A single, more basic cause may result in multiple failures which fail the system. For example, an operator may have miscalibrated all the sensors, or one steam line break may cause all instrumentation to fail in a control panel. Multiple failures which can fail the system and which can originate from a common cause are termed common cause failures.

In evaluating a fault tree, we do not know which failures will be common cause failures; however, we can indicate the susceptibility that component failures may have to a common initiating cause. Now by definition, the top event occurs, i.e., system failure occurs, if all the primary failures in a minimal cut set occur. Therefore, we are interested only in those common causes which can trigger all the primary failures in a minimal cut set. A cause which does not trigger all the primary failures in a minimal cut set will not by itself cause system failure.

To identify minimal cut sets which are susceptible to common cause failures we can first define common cause categories, which are general areas that can cause component dependence. Examples of common cause categories include manufacturer, environment, energy sources (not explicitly shown in the tree), and humans. The list below gives some example categories which might be considered in a common cause susceptibility evaluation.

List of Common Cause Categories to be Evaluated

Manufacturer
Location
Seismic Susceptibility
Flood Susceptibility
Temperature
Humidity
Radiation
Wear-out Susceptibility
Test Degradation
Maintenance Degradation
Operator Interactions
Energy Sources
Dirt or Contamination

For each common cause category, we then define specific "elements." For example, for the category "Manufacturer" the elements would be the particular manufacturers involved which we might code as "Manufacturer 1." "Manufacturer 2," etc. For the "Location" category, we might divide the plant into a given number of physical locations which would be the elements. For the category "Seismic Susceptibility," we might define several sensitivity levels ranging from no sensitivity to extreme sensitivity, and for more specificity we would define acceleration ranges where failure would likely occur.

Our next task in the common cause susceptibility evaluations involves component coding. As part of the component name code or in associated component description fields, for each component failure we denote the element of each category associated with the component. The categories and elements can be indexed or keyed according to any convenient coding system. For example, "MV2-183" may denote manual valve 2 ("MV2") having element 1 associated with category 1, having element 8 associated with category 2, and element 3 associated with category 3 ("-183"). This kind of naming can easily be coded on the fault tree and in any subsequent computer input.

Having performed this coding, we can then identify the potentially susceptible minimal cut sets among the collection of minimal cut sets determined for the fault tree. The minimal cut sets which are potentially susceptible to common cause failures are those whose primary failures all have the same element of a given category. Having identified the potentially susceptible minimal cut sets, we then need to finally screen these cut sets to determine those which may require further action. This final screening may be based on past histories of common cause occurrences, some sort of quantification analysis, and/or engineering judgement. This last step is the most difficult and most time consuming. Chapter XII, section 6, discusses computer codes which can perform the initial searches.

3. Quantitative Evaluations

Once the minimal cut sets are obtained, probability evaluations can be performed if quantitative results are desired. The quantitative evaluations are most easily performed in a sequential manner, first determining the component failure probabilities, then the minimal cut set probabilities, and finally the system, i.e., top event, probability. Quantitative measures of the importance of each cut set and of each component can also be obtained in this process.

If the failure rates are treated as random variables, then random variable propagation techniques can be used to estimate the variabilities in system results which result from the failure rate variations. We first discuss the usual "point estimations" where one value is assigned to each failure rate and where one value is obtained for each minimal cut set probability and for the system probability. Afterwards, we will discuss random variable analyses.

(a) Component Failure Probability Models

By "component" we mean any basic primary event shown on the fault tree (circle, diamond, etc.). For any component, we consider only failure probability models for which either a constant failure rate per hour applies or a constant failure rate per cycle applies. In using these constant failure rate models, we ignore any time-dependent effects such as component burn-in and component wear-out. The constant failure rate models which we discuss are usually used for order of magnitude results. Where time-dependent effects such as burn-in or wear-out are important or where precision better than, say, a factor of 10 is required, then more sophisticated models are required. These more sophisticated models include, for example, Weibull and Gamma failure distribution modeling; the reader is referred to [12] and [17] for discussions of these other models.

(b) Constant Failure Rate Per Hour Model: Probability Distributions

Consider first a component whose failures are modeled as having a constant failure rate per hour. Let the constant failure rate per hour be denoted by λ. When we use the constant failure rate per hour model, which we shall simply call the λ-model, then we are assuming that failure probabilities are directly related to component exposure times. The longer the exposure time period, the higher the probability of failure. The failure causes can be human error, test and maintenance, or environmental, such as contamination, corrosion, etc. The λ-model is the model most often used in fault tree quantifications.

For the λ-model, the resulting first failure probability distributions are exponential distributions. We shall review the distribution properties to refresh the reader's memory. The probability F(t) that the component suffers its first failure within time period t, given it is initially working, is:

$$F(t) = 1 - e^{-\lambda t}. \tag{XI-1}$$

The quantity F(t) is the cumulative probability distribution as discussed in section 3 of Chapter X. In reliability terminology, F(t) is called the component unreliability. The complementary quantity to F(t) is the quantity $1 - F(t)$, which is the probability of no failure in time period t (given the component is initially working):

$$1 - F(t) = e^{-\lambda t}. \tag{XI-2}$$

In statistical terminology, the quantity $1 - F(t)$ is called the complementary cumulative probability. In reliability terminology, $1 - F(t)$ is the component reliability and is denoted by R(t):

$$R(t) = 1 - F(t). \tag{XI-3}$$

The density function denoted by f(t) is the derivative of F(t); the quantity f(t)Δt, where the interval Δt approaches zero, is the probability that the component does not fail in time period t but then does fail in some small interval Δt about t. As part of the definition of f(t)Δt, we again assume the component is initially working at the beginning of the period. This "initially working" assumption is applied to all the calculations, and we will not explicitly state it in the forthcoming discussions.

For the exponential distribution, the density function f(t) is:

$$f(t) = \lambda e^{-\lambda t}. \tag{XI-4}$$

The constant failure rate per hour model is so named because a formal calculation of the time-dependent failure rate $\lambda(t)$ simply gives the constant value λ. Estimates of the constant failure rate λ are given for a variety of components in various data sources. The analyst needs to obtain a value of λ for each component failure on his fault tree for which the constant failure rate model is applied. Table XI-1 gives some representative failure rates for various component failures; the data are taken from WASH-1400 (reference [38]).

Table XI-1. Representative Failure Rates from WASH-1400

		ASSESSMENT -MEDIAN-	LOWER & UPPER BOUND			ASSESSMENT -MEDIAN-	LOWER & UPPER BOUND
	Failure to Operate	3×10^{-4}/D	$1 \times 10^{-4} - 1 \times 10^{-3}$		Failure To Start	1×10^{-3}/D	3×10^{-4} 3×10^{-3}
CLUTCH ELEC	Premature Open	1×10^{-6} HR	$1 \times 10^{-7} - 1 \times 10^{-5}$	Pumps	Failure To Run —Normal	3×10^{-5}/HR	3×10^{-6}·3×10^{-4}
CLUTCH MECH	Failure to Open	3×10^{-7}/HR	$3 \times 10^{-8} - 3 \times 10^{-6}$		Failure To Run —Extreme ENV	1×10^{-3}/HR	1×10^{-4}·1×10^{-2}
	Failure to Operate	3×10^{-4}/D	$1 \times 10^{-4} \cdot 1 \times 10^{-3}$		Fails To Operate	1×10^{-3}/D	3×10^{-4}·3×10^{-3}
SCRAM RODS	Failure to Insert (Single Rod)	1×10^{-4}/D	$3 \times 10^{-5} - 3 \times 10^{-4}$	Valves: MOV.	(Plug) Failure To Remain Open	1×10^{-4}/D	3×10^{-5}·3×10^{-4}
ELECTRIC MOTORS	Failure to Start	3×10^{-4}/D	1×10^{-4} 1×10^{-3}		External Leak Or Rupture	1×10^{-8}/HR	1×10^{-9} 1×10^{-7}
	Failure to Run	1×10^{-5}/HR	3×10^{-6} 3×10^{-5}	Valves (SOV)	Fails To Operate	1×10^{-3}/D	3×10^{-4}·3×10^{-3}
	Failure to Run (Extreme ENVIR)	1×10^{-3}/HR	1×10^{-4} 1×10^{-2}		Fails To Operate	3×10^{-4}/D	1×10^{-4}·1×10^{-3}
	Failure to Energize	1×10^{-4}/D	3×10^{-5} 3×10^{-4}	Valves (AOV)	(Plug) Failure To Remain Open	1×10^{-4}/D	3×10^{-5}·3×10^{-4}
RELAYS	Failure NO Contact to Close	3×10^{-7} HR	1×10^{-7} 1×10^{-6}		External Leak—Rupture	1×10^{-8}/HR	1×10^{-9} 1×10^{-7}
	Short Across NO/NC Contact	1×10^{-8} HR	1×10^{-9} 1×10^{-7}		Failure To Open	1×10^{-4}/D	3×10^{-5} 3×10^{-4}
	Open NC Contact	1×10^{-7} HR	3×10^{-8} 3×10^{-7}	Valves (Check)	Reverse Leak	3×10^{-7}/HR	1×10^{-7} 1×10^{-6}
SWITCHES	Limit: Failure to Operate	3×10^{-4}/D	1×10^{-4} 1×10^{-3}		External Leak—Rupture	1×10^{-8}/HR	1×10^{-9} 1×10^{-7}
	Torque: Fail to OPER	1×10^{-4} D	3×10^{-5}·3×10^{-4}	Valves (Vacuum)	Failure To Operate	3×10^{-5}/D	1×10^{-5} 1×10^{-4}
	Pressure Fail to OPER	1×10^{-4} D	3×10^{-5}··3×10^{-4}		Rupture	1×10^{-8}/HR	1×10^{-9} 1×10^{-7}
	Manual, Fail to TRANS	1×10^{-5}/D	3×10^{-6} 3×10^{-5}	Valves: Orifices, Flow Meters, (Test)	Rupture	1×10^{-8}/HR	1×10^{-9} 1×10^{-7}
	Contacts Short	1×10^{-8} HR	1×10^{-9}·1×10^{-7}	Valves (Manual)	Failure To Remain Open (Plug)	1×10^{-4}/D	3×10^{-5} 3×10^{-4}
CIRCUIT BREAKERS	Failure to Operate	1×10^{-3} D	$3 \times 10^{-4} \cdot 3 \times 10^{-3}$	Valves (Relief)	Fail To Open/D	1×10^{-5}/D	3×10^{-6} 3×10^{-5}
	Premature Transfer	1×10^{-6}/HR	$3 \times 10^{-7} - 3 \times 10^{-6}$		Premature Open/HR	1×10^{-5}/HR	3×10^{-6}·3×10^{-5}
FUSES	Premature, Open	1×10^{-6}/HR	$3 \times 10^{-7} - 3 \times 10^{-6}$	Pipes> 3" HI Quality	Rupture (Section)	1×10^{-10}/HR	3×10^{-12} 3×10^{-9}
	Failure to Open	1×10^{-5}/D	$3 \times 10^{-6} - 3 \times 10^{-5}$	Pipes< 3"	Rupture	1×10^{-9}/HR	3×10^{-11} 3×10^{-8}
WIRES	Open	3×10^{-6}/HR	1×10^{-6} 1×10^{-5}	Gaskets	Leak	3×10^{-6}/HR	1×10^{-7} 1×10^{-4}
	Short to GND	3×10^{-7}/HR	3×10^{-8}·3×10^{-6}	Flanges, Closures, Elbows:	Leak/Rupture	3×10^{-7}/HR	1×10^{-8}·1×10^{-5}
	Short to PWR	1×10^{-8} HR	1×10^{-9}·1×10^{-7}	Welds	Leak	3×10^{-9}/HR	1×10^{-10} 1×10^{-7}
TRANSFORMERS	Open CKT	1×10^{-6}/HR	3×10^{-7}·3×10^{-6}	Diesel (Complete Plant) (Emergency Loads) Diesel (Engine Only)	Failure to Start Od	3×10^{-2}/D	1×10^{-2}·1×10^{-1}
	Short	1×10^{-6} HR	3×10^{-7}·3×10^{-6}		Failure to Run λo	3×10^{-3}/HR	3×10^{-4}·3×10^{-2}
SOLID STATE DEVICES	Fails to Function HI PWR Application	3×10^{-6}/HR	3×10^{-7}·3×10^{-5}		Failure to Run λo	3×10^{-4}/HR	3×10^{-5}·3×10^{-3}
	Shorts	1×10^{-6}/HR	$1 \times 10^{-7} - 1 \times 10^{-5}$	Batteries Power Supplies	NO/Output λs	3×10^{-6}/HR	1×10^{-6}·1×10^{-5}
	Fails To Function Low PWR Application	1×10^{-6}/HR	$1 \times 10^{-7} - 1 \times 10^{-5}$	Instrumentation (Amplifiers, Annunciators, Transducers, Combination)	Failure to Operate, λs	1×10^{-6}/HR	1×10^{-7}·1×10^{-5}
	Shorts	1×10^{-7}/HR	$1 \times 10^{-8} - 1 \times 10^{-6}$		Shift Calibration λo	3×10^{-5}/HR	3×10^{-6}·3×10^{-4}

FAILURE MODES

Extreme precision is not required (and is not believed!) in a fault tree evaluation; it is the <u>order of magnitude size</u> of the failure rate that is of concern, i.e., is the failure rate 10^{-6} per hour or 10^{-5} per hour. To this "order of magnitude" precision, detailed environments and detailed component specifications are often not important in obtaining gross estimates of the failure rate. The analyst, however, should of course use all the available information in obtaining as precise an estimate as he can for λ for each component or basic event on his tree.

Because extreme precision is not required in a fault tree evaluation, the exponential distribution can be approximated by its first order term to simplify the calculations. The cumulative exponential distribution, i.e., the component unreliability, is thus approximated as:

$$F(t) \cong \lambda t. \tag{XI-5}$$

The above approximation is accurate to within 5% for failure probabilities ($F(t)$) less than 0.1 and the slight error made is on the conservative side. Furthermore, the approximation error is small compared to uncertainties in λ.

The failure rate λ used in the λ-model can be either a <u>standby failure rate</u> or an <u>operating failure rate</u>; data sources give both types of rates. If λ is a standby failure rate, then the time period t used in Equation (XI-5) should be the standby time t, i.e., the time period during which the component is "ready" for actual operation. For these standby situations, $F(t)$ is the probability that the failure will occur in standby. If λ is an operating failure rate, then t is the actual operating time period and $F(t)$ is the probability that the failure will occur in operation. Many components will have both a standby failure rate and an operating failure rate; for example, a pump will have a standby failure rate when it is not operating and will have an operating failure rate when it is operating. The analyst must ensure that the proper failure rate is used with the proper time period. For a standby and an operating phase the total failure probability is

$$F_s(t_s) + (1 - F_s(t_s))\, F_o(t_o) \cong F_s(t_s) + F_o(t_o),$$

where the subscript "s" denotes the standby phase and the subscript "o" the operating phase. For small probabilities (less than, say, 0.1) we can thus simply sum the failure probabilities.

(c) The Constant Failure Rate Per Hour Model: Reliability Characteristics

As stated in the previous section, the quantity $R(t) = 1 - F(t)$ is the probability of no failure in time t and is termed the component reliability. The component unreliability $F(t)$ is the probability of at least one failure in time period t. It is also the probability of first failure in time t. These definitions of $F(t)$ incorporate the possibility of more than one failure occurring if the component failures are repairable (the reason for the phrase "at least one failure"). If the component failure is not repairable, then at most one failure will occur.

When we say failures are repairable, we mean that the component is repaired or replaced when it fails. The repair or replacement need not begin immediately after failure, and when begun will require some time to complete. The repair or replacement operation can be characterized by the downtime of the component denoted by d, which is the total period of time the component is down and unavailable for operation. For a standby component d is the downtime during which a demand on the component may occur. If the plant is shutdown sometime after the failure occurs then d is only that period of online downtime in which the component could still be required to operate (to respond to, say, an accident).

The cumulative distribution for the downtime denoted by G(d) is defined as follows:

$$G(d) = \text{the probability that the downtime period will be less than d.} \qquad \text{(XI-6)}$$

The cumulative distribution G(d) is obtained from experience data on repairs/ replacements and completely defines the repair/replacement process for quantitative evaluations.

Let q(t) be the <u>component unavailability</u> and be defined as:

$$q(t) = \text{the probability that the component is down at time t}$$
$$\text{and unable to operate if called on.} \qquad \text{(XI-7)}$$

$1 - q(t)$ is the <u>component availability</u> and is the probability that the component is up and able to operate were it called on.

If component failures are not repairable, then the component will be down at time t if and only if it has failed in time t. Consequently, for nonrepairable failures, where the component is up at $t = 0$, the unavailability q(t) is equal to the unreliability F(t):

$$q(t) = F(t); \text{nonrepairable failures.} \qquad \text{(XI-8)}$$

For the exponential distribution, the unavailability q(t) is thus simply given by the approximation:

$$q(t) \cong \lambda t. \qquad \text{(XI-9)}$$

For nonrepairable failures, the constant failure rate λ is thus all that is needed to calculate the basic component characteristics F(t) and q(t) which are used in a fault tree evaluation.

For repairable failures, the component unavailability q(t) is not equal to the unreliability and we need information on the repair process* to calculate q(t).

We will assume that the repair restores the component to a state where it is essentially as good as new. This assumption is optimistic but is usually made. The effects of test inefficiencies can be investigated by more sophisticated analyses. (For other treatments see references [37] and [41].)

---

*From here on, we will speak only of "repair" but we will mean repair or replacement.

For repairable failures, we consider two cases: (1) when failures are monitored, and (2) when failures are not detectable until a periodic surveillance test is performed. For the monitored case, when a failure occurs an alarm, annunciator, light, or some other signal alerts the operator. In this case, the unavailability $q(t)$ quickly reaches a constant asymptotic value q_M which is given by:

$$q_M = \frac{\lambda T_D}{1 + \lambda T_D} \qquad\qquad\qquad (XI\text{-}10)$$

$$\cong \lambda T_D. \qquad\qquad\qquad (XI\text{-}11)$$

The failure rate λ is the standby failure rate and the quantity T_D is the average online downtime obtained by statistically averaging the downtime distribution (described by the cumulative distribution $G(d)$). The downtime which is evaluated is again the online downtime during which the system is up and the component may be called on (for example, to respond to an accident situation). For simplified evaluations, the downtimes can often be broken into several discrete values with associated probabilities and a statistical average taken over the discrete values. The approximation given by Equation (XI-11) is conservative and is within 10% accuracy for $\lambda T_D < 0.1$.

For components which are not monitored but which are periodically tested, any failures occurring are not detectable until the test is performed. This is the situation, for example, when surveillance tests are performed monthly; any failure which occurred during the past month would be detected only when the test is performed. (We assume perfect testing here, in that essentially 100% of the failure modes are detected.)

For periodic tests performed at intervals of T, the unavailability rises from a low of $q(t = 0) = 0$ immediately after a test is performed to a high value of $q(t = T) = 1 - e^{-\lambda T} \cong \lambda T$ immediately before the next test is performed. Because the exponential can be approximated by a linear function (for $\lambda T < 0.1$ say) the average unavailability between tests is approximately $\lambda T/2$. The average value is applicable for fault tree evaluations if we assume that a demand on the component may occur uniformly at any time in the interval.

If the component is found failed at a surveillance test, then it will remain down during the necessary repair time. Considering this additional repair contribution, we have the following expression for the total average unavailability q_T for periodically tested components:

$$q_T = \lambda T/2 + \lambda T_R. \qquad\qquad\qquad (XI\text{-}12)$$

In the above formula, λ is again the standby failure rate per hour and T_R is the average repair time obtained from downtime considerations. The repair time evaluated is again the online repair time during which the component may be called on to function. The subscript "R" is used on T_R to denote that it is the average repair time and not the total downtime which is the sum of the repair time plus the undetected downtime from time of failure to time of detection.

In general, T_R is small compared to T and the second term on the right hand side of Equation (XI-12) is negligible; we thus have:

$$q_T \cong \lambda T/2, T_R << T. \qquad \text{(XI-13)}$$

For repairable failures, the unavailability is thus given by q_M or q_T depending on whether monitoring exists or periodic testing is performed with no monitoring between tests. (If monitoring is performed, q_M applies regardless of whether any additional periodic testing is performed.) For each repairable component of the tree, λ and T_D (monitored) or λ, T_R, and T (periodically tested) are required as data inputs. Failure rate data sources supply λ and operating specifications for the component are sources for T_R, T, and T_D.

In addition to the component unavailability, there is one component <u>reliability</u> characteristic which is of interest when an operational system is being evaluated. This component characteristic is the <u>component failure occurrence rate</u> w(t) and is defined such that:

$w(t)\Delta t$ = the probability that the component fails in time
   t to t + Δt. $\qquad \text{(XI-14)}$

In the definition of w(t), we are not given that the component has operated without failure to time t as was the case for the failure rate definition $\lambda(t)$ (see Section 8 of Chapter X). In fact, if the component is repairable, then it can have failed many times previously; the quantity $w(t)\Delta t$ is the probability that it fails in time t to t + Δt irrespective of history.

The occurrence rate w(t) is applicable for both nonrepairable and repairable components. For both repairable and nonrepairable components, the expected number of failures in some time interval t_1 to t_2, denoted by $n(t_1, t_2)$, is given by the integral of w(t) from t_1 to t_2:

$$n(t_1, t_2) = \int_{t_1}^{t_2} w(t)dt. \qquad \text{(XI-15)}$$

For <u>nonrepairable</u> component failures, the component can only fail once. Therefore, w(t) is equal to the probability density function for first failure:

$$w(t) = f(t) \qquad \text{(XI-16)}$$

$$= \lambda e^{-\lambda t} \qquad \text{(XI-17)}$$

where Equation (XI-17) is for the constant failure rate model (the λ-model).

For time t small compared to $1/\lambda$ (such that $\lambda t < 0.1$), $e^{-\lambda t}$ is approximately 1, and hence Equation (XI-17) becomes,

$$w(t) \cong \lambda, \lambda t < .1. \qquad \text{(XI-18)}$$

For repairable failures, $w(t)$ can be a complex function of time; however, it approaches λ as time progresses, and this asymptotic value of λ is generally precise enough for applications:

$$w(t) \cong \lambda. \tag{XI-19}$$

Hence for both nonrepairable and repairable failures, $w(t) = \lambda$ is generally a reasonable approximation. (Some of the computer codes discussed in the next chapter can compute the time dependent values of $w(t)$.)

(d) Reliability Characteristics for the Constant Failure Rate Per Cycle Model

Instead of modeling component failures as having a constant failure rate per hour, we can use a constant failure per cycle model. In the constant failure rate per cycle model, the component is assumed to have a constant probability of failing when it is called on (i.e., when it is "cycled"). This probability of failing per cycle, which we denote by p, is independent of any exposure time interval, such as the time interval between the test or the time that the component has existed in standby.

The constant failure rate per cycle model, which we shall simply call the p-model, is applied when failures are inherent to the component and are not caused by "external" mechanisms which are associated with exposure time. For cyclic failures, the cycling of the component may actually cause the failure (because of stress, etc.). For example, a component which is obtained from a manufacturer and immediately placed in the field may be modeled as having a certain failure probability p due to manufacturing defects. After pre-operational testing (i.e., burn-in testing), many of the inherent component failures would be detected and the failures might then be best modeled by the λ-model, i.e., the constant failure rate per hour model.

In past practice, the p-model has been applied to relatively few components, whereas the λ-model, the constant failure rate per hour model, has been applied to the majority of components. The analyst must decide which component model, the p-model or the λ-model, is most applicable for his analyses. Failure rate data sources sometimes indicate the appropriate models; otherwise, the analyst must decide based on knowledge of the failure causes and mechanisms.

The reliability characteristics for the p-model are straightforward and are all based on the one characterizing value p, which is the probability of failure per cycle, or per demand. We again use the definition of component unreliability, $F(t)$, and component unavailability, $q(t)$, as given by Equations (XI-8) through (XI-11). For n demands in time t and assuming independent failures, the reliability (R_c) and the unavailability (q_c) are given by:

$$R_c = 1 - q_c = (1 - p)^n \tag{XI-20}$$

or

$$1 - R_c = q_c \cong np, \, np < 0.1. \tag{XI-21}$$

(The subscript c in R_c and q_c denotes "cyclic" values.)

As noted in the above equations, the reliability and unavailability do not depend explicitly on time but on the number of cycles (demands) occurring in that time. For one demand (n = 1), we note that $1 - R_c = q_c = p$. For each component on the fault tree modeled by the p-model, the user must obtain the appropriate value of p and the number of demands (usually one).

(e) Minimal Cut Set Reliability Characteristics

Once the component reliability characteristics are obtained, the reliability characteristics for the minimal cut sets can be evaluated. For a fault tree of a standby system, such as a nuclear safety system, the characteristic of principal concern is the minimal cut set unavailability denoted by Q:

> Q(t) = the probability that all the components in the minimal
>   cut set are down at time t and unable to operate.   (XI-22)

Because a minimal cut set can be viewed as being a particular failure mode of the system we can also define Q as:

> Q(t) = the probability that the system is down at time t
>   due to the particular minimal cut set.   (XI-23)

We can thus also call Q(t) the system unavailability due to a minimal cut set.

We can index each minimal cut set of the fault tree in any way we choose, and then $Q_i(t)$ would be the unavailability for minimal cut set i. To determine $Q_i(t)$ we note that, by its definition, the minimal cut set is an intersection of the associated component failures; the minimal cut set occurs if and only if all the component failures occur. Assuming the component failures are independent, recall from Chapter VII (Equation VII-3) that the probability of an intersection (i.e., an AND gate) is simply the product of the component probabilities. Hence:

$$Q_i(t) = q_1(t)q_2(t) \ldots q_{n_i}(t)$$ (XI-24)

where $q_1(t)$, $q_2(t)$, etc., are the unavailabilities of the component in the particular minimal cut set and n_i is the number of components in the cut set. As an example in applying Equation (XI-24), if a minimal cut set has two components, having respective unavailabilities of 1×10^{-2} and 1×10^{-3}, then the cut set unavailability is

$$Q_i = (1 \times 10^{-2})(1 \times 10^{-3}) = 1 \times 10^{-5}.$$

The component unavailabilities are those given in the previous sections; any combinations of component unavailabilities can be used (e.g., one component can be periodically tested and the other can have a cyclic failure rate, etc.). If the components in the minimal cut set are all repairable or cyclic, then constant values can be used for the component unavailabilities, e.g., (Equation (XI-10) or (XI-11)), in which we ignore any time-dependent transient behaviors. For this completely repairable or cyclic case and within our approximations, the minimal cut set unavailability is then simply a constant and independent of time.

If the fault tree is of an operational system, then instead of the unavailability, the number of system failures and the probability of no system failure are of primary interest. A minimal cut set characteristic giving information on these reliability-related concerns and one which is easily calculable is the <u>minimal cut set occurrence rate</u> denoted by $W(t)$. The minimal cut set occurrence rate $W(t)$ is defined such that:

$$W(t)\Delta t = \text{the probability that the minimal cut set}$$
$$\text{failure occurs in time } t \text{ to time } t + \Delta t. \qquad \text{(XI-25)}$$

The quantity Δt is a small increment of time. The occurrence rate itself, $W(t)$, is thus a probability per unit time of the minimal cut set failure occurring. Because a minimal cut set can be considered a system failure mode we can equivalently define $W(t)\Delta t$ to be:

$$W(t)\Delta t = \text{the probability that a system failure occurs in time}$$
$$t \text{ to } t + \Delta t \text{ by the particular minimal cut set.} \qquad \text{(XI-26)}$$

If we index all the minimal cut sets of the tree, then $W_i(t)$ refers to the occurrence rate of minimal cut set i.

To calculate $W_i(t)$, we use the basic definition of a minimal cut set and the concept of an "occurrence." A minimal cut set failure occurs at time t to t + Δt if all the components except one are down at time t and the other component then fails at time t to t + Δt. Assuming independence of the component failures $W_i(t)$ is thus given by:

$$\begin{aligned}
W_i(t)\Delta t = &\, q_2(t)q_3(t)\ldots q_{n_i}(t)w_1(t)\Delta t \\
&+ q_1(t)q_3(t)\ldots q_{n_i}(t)w_2(t)\Delta t \\
&+ q_1(t)q_2(t)\ldots q_{n_i}(t)w_3(t)\Delta t \\
&+ \\
&\quad \cdot \\
&\quad \cdot \\
&\quad \cdot \\
&+ q_1(t)q_2(t)\ldots q_{n_i-1}(t)w_{n_i-1}(t)\Delta t
\end{aligned} \qquad \text{(XI-27)}$$

where the $q(t)$'s are the component unavailabilities and the $w(t)$'s are the component occurrence rates (Equation (XI-14)). The first term on the right hand side of Equation (XI-27) is the probability that all components except component 1 are down at time t and then component 1 fails. The second term is the probability that component 2 is the component that fails in time t to t + Δt, all other components being already down, etc. The contribution from each of the components failing in time t to t + Δt is summed over the n_i components in the cut set to obtain the total occurrence rate given by Equation (XI-27). The Δt's cancel in Equation (XI-27), giving:

$$
\begin{aligned}
W_i(t) = {} & q_2(t)q_3(t)\ldots q_{n_i}(t)w_1(t) \\
& + q_1(t)q_3(t)\ldots q_{n_i}(t)w_2(t) \\
& + q_1(t)q_2(t)\ldots q_{n_i}(t)w_3(t) \\
& + \\
& \quad \cdot \\
& \quad \cdot \\
& \quad \cdot \\
& + q_1(t)q_2(t)\ldots q_{n_i-1}(t)w_{n_i}(t).
\end{aligned}
\tag{XI-28}
$$

The minimal cut set occurrence rate $W_i(t)$ is strictly applicable when all the components have a per hour failure rate (the λ-model). Cyclic components (the p-model), as previously stated, do not have any explicit time-associated behaviors. The above equation can be applied to minimal cut sets having cyclic components, if we use for the cyclic components $q_c(t) \cong n(t)p$ and $w_c(t) \cong pk(t)$, where p is the cyclic component failure probability, $n(t)$ is the (expected) number of demands in time t, and $k(t)$ is the probability of a demand occurring per unit time at time t. (The quantity $k(t)\Delta t$ is thus the probability that a demand occurs in time t to $t + \Delta t$.) The quantities $n(t)$ and $k(t)$ must be obtained from operational considerations for the component.

The expected number $N_i(t_1,t_2)$ of failures of minimal cut set i occurring during some time period from t_1 to t_2 is:

$$
N_i(t_1,t_2) = \int_{t_1}^{t_2} W_i(t)dt.
\tag{XI-29}
$$

If the components in the minimal cut set are all repairable, and constant values are used for the component unavailabilities and occurrence rates (ignoring any transients), then $W_i(t)$ is a constant, $W_i(t) = W_i$. In this constant value case $N_i(t_1,t_2)$ is simply equal to the time interval multiplied by the constant minimal cut set occurrence rate:

$$
N_i(t_1,t_2) = (t_2 - t_1)W_i.
\tag{XI-30}
$$

Because the system fails every time the minimal cut set fails (by the minimal cut set definition), $N_i(t_1,t_2)$ is the expected number of times the system fails in time period t_1 to t_2, due to minimal cut set i. When all components are nonrepairable then $N_i(t_1,t_2)$ is also the probability that the minimal cut set fails in the time period t_1 to t_2 (the expected number being equal to the probability in this case). Even when components are repairable, if the system failure probability over the time period of interest is small (say, less than 0.1), then $N_i(t_1,t_2)$ is less than 1. Even though $N_i(t_1,t_2)$ is strictly the expected number of failures, it is also a reasonably good approximation for the probability of the minimal cut set failing in time period t_1 to t_2. This approximation in the repairable case is conservative (the true probability is less than $N_i(t_1,t_2)$) and fairly accurate, generally agreeing to within 10% of the true probability for $N_i(t_1,t_2) < 0.1$.

Using our previous nomenclature, the probability of minimal cut set failure is the minimal cut set unreliability and hence, using the above reasoning, when $N_i(t_1,t_2) < 0.1$:

$$N_i(t_1,t_2) \cong \text{minimal cut set unreliability in time } t_1 \text{ to } t_2. \tag{XI-31}$$

In terms of system failure, when $N_i(t_1,t_2) < 0.1$, then $N_i(t_1,t_2)$ is thus also approximately the probability of system failure in time t_1 to t_2 due to minimal cut set i. We can thus say, within our approximation, that $N_i(t_1,t_2)$ is the system unreliability due to minimal cut set i. When some components are repairable, the exact minimal cut set and system unreliabilities are generally difficult to calculate and $N_i(t_1,t_2)$ consequently provides a useful and generally accurate approximation which is simple to calculate and good enough for most applications.

For a fault tree evaluation, the minimal cut set unavailabilities, $Q_i(t)$, and the minimal cut set occurrence rates, $W_i(t)$, provide comprehensive information on the probabilistic behavior of the minimal cut sets. If a standby system, such as a nuclear safety system, is being evaluated, then only the cut set unavailability $Q_i(t)$ is usually calculated. The minimal cut set characteristics need to be calculated for all the dominant minimal cut sets of the tree. For small numbers of minimal cut sets, the characteristics can be calculated for all the cut sets of the fault tree. For fault trees having very large numbers of minimal cut sets, the minimal cut set characteristics are generally only computed for the lower order cut sets, e.g., single-component and double-component cut sets. Because component failures are assumed to be independent, the values of $Q_i(t)$ and $W_i(t)$ for higher order cut sets (e.g., triples and higher) are generally negligible compared to those for the lower order cut sets (singles and doubles). The independence assumption represents an optimal condition and the true cut set characteristics, for doubles and up, may be quite higher than the calculated values of $Q_i(t)$ and $W_i(t)$ because of dependencies among the component failures. If the fault tree has only double cut sets and higher, then the calculated values for $Q_i(t)$ and $W_i(t)$ represent design capability numbers useful principally for relative evaluations; the actual achieved values of $Q_i(t)$ and $W_i(t)$ may be quite higher and will be much more difficult to estimate. (See for example reference [12] for further considerations.)

(f)   System (Top Event) Reliability Characteristics

Once the minimal cut set characteristics are obtained, the determination of the system characteristics is quite straightforward. The system unavailability denoted by $Q_s(t)$ (the subscript "s" denoting "system") is defined as:

$$Q_s(t) = \text{the probability that the system is down at time t and unable to operate if called on.} \tag{XI-32}$$

For a standby system, such as a nuclear safety system, $Q_s(t)$ is the most critical system characteristic. If the top event of the fault tree is not a system failure but some general event, then $Q_s(t)$ is the probability that the top event exists at time t (having occurred earlier and persisting to time t).

Now, the system is down if and only if any one or more of the minimal cut sets is down. If we ignore the possibility of two or more minimal cut sets being simultaneously down, the system unavailability $Q_s(t)$ can thus be approximated as the sum of the minimal cut set unavailabilities $Q_i(t)$:

$$Q_s(t) \cong \sum_{i=1}^{N} Q_i(t) \tag{XI-33}$$

where Σ denotes a summation of $Q_i(t)$ over the N minimal cut sets considered in the tree.

Equation (XI-33), the so-called "rare event approximation," was first introduced in Chapter VI. It generally gives results agreeing within 10% of the true unavailability for $Q_s(t) < 0.1$. Furthermore, any error made is on the conservative side in that the true unavailability is slightly lower than that computed by Equation (XI-33). Equation (XI-33) is usually used in fault tree evaluations; it is simple to calculate, and it can be truncated at any value of N to consider only those cut sets contributing most to $Q_s(t)$. If the component failures are all repairable or cyclic and constant values used for the component unavailabilities, then $Q_s(t)$ is independent of time and is simply a constant value Q_s.

For on-line operational systems, the <u>system failure occurrence rate</u> $W_s(t)$ is of interest and is defined such that:

$$W_s(t)\Delta t = \text{the probability that the system fails in time}$$
$$t \text{ to } t + \Delta t. \tag{XI-34}$$

The occurrence rate itself, $W_s(t)$, is the probability per unit time of system failure at time t. (For any general top event, $W_s(t)$ is the probability per unit time that the top event occurs at time t.)

The system failure occurs if and only if any one or more of the minimal cut sets occurs. The system failure occurrence rate then, $W_s(t)$, can be expressed as the sum of the minimal cut set occurrence rates $W_i(t)$:

$$W_s(t) = \sum_{i=1}^{N} W_i(t). \tag{XI-35}$$

Equation (XI-35) is another application of the "rare event approximation." It is quite accurate for low probability events because for such events, the probability of two or more minimal cut sets occurring simultaneously is negligible. Equation (XI-35) is again simple to evaluate and it can be truncated so that only the N dominant minimal cut contributors are considered.

XI-20   **FAULT TREE HANDBOOK**

If we use $W_s(t)$, the expected number of system failures $N_s(t_1,t_2)$ in time t_1 to t_2 is:

$$N_s(t_1,t_2) = \int_{t_1}^{t_2} W_s(t)dt. \qquad \text{(XI-36)}$$

As a particular use of the above formula, the expected number of system failures in time t, $N_s(t)$, is:

$$N_s(0,t) = \int_0^t W_s(t')dt'. \qquad \text{(XI-37)}$$

If the component failures are all repairable or cyclic and constant values used for the component unavailabilities, then $W_s(t)$ is simply a constant W_s, and $N_s(t_1 t_2)$ is simply W_s times the interval $t_2 - t_1$.

Using the same rationale as for the minimal cut set $N_i(t_1,t_2)$, for $N_s(t_1,t_2)$ less than 0.1, $N_s(t_1,t_2)$ is also a reasonably accurate approximation for the probability of system failure in time t_1 to t_2, which is the system unreliability:

$$N_s(t_1,t_2) \cong \text{system unreliability in time } t_1 \text{ to } t_2. \qquad \text{(XI-38)}$$

The quantity $N_s(0,t)$ is thus a reasonably accurate approximation of the system unreliability in time period t.

The system unavailability $Q_s(t)$, the system failure occurrence rate $W_s(t)$, and the expected number of system failures $N_s(t_1,t_2)$ give comprehensive information on the probabilistic description of system failure. In using these results, the reader must keep in mind the assumptions and limitations of the calculations, particularly the assumption of independence of component failure occurrences. As discussed for the minimal cut set characteristics (section e), if the fault tree has only double cut sets and higher, the system results as calculated here may be very much below the true values due to dependencies among the component failures. When these dependencies exist to the extent that they significantly increase failure probabilities, then $Q_s(t)$, $W_s(t)$ and $N_s(t_1,t_s)$ represent optimal design-based numbers which are useful for relative evaluations but are not useful for absolute evaluations.

(g) Minimal Cut Set and Component Importances

As an additional evaluation we describe a quantitative technique for determining the "importance" of each minimal cut set and each component failure. We define the minimal cut set importance to be the fraction of system failure probability contributed by a particular minimal cut set. We define the component importance to be the fraction of system failure probability contributed by the particular component failure. Different formulas can be used to calculate the importances (see reference [21] for a discussion of the different approaches); for our discussion here we will use one of the simplest methods of calculating the importances.

The minimal cut set importance and the component importance can be calculated with regard to the system unavailability, $Q_s(t)$, or the system failure occurrence rate,

$W_s(t)$. The rule in either case is the same: to calculate the minimal cut set importance, we take the ratio of the minimal cut set characteristic over the system characteristic. For the component importance, we sum the characteristic of all minimal cut sets containing the component and divide by the system characteristic.

Let $E_i(t)$ be the importance of minimal cut set i at time t, and let $e_k(t)$ be the importance of component k at time t (we have simply indexed the cut sets and components for ease of identification). With regard to system unavailability, then:

$$E_i(t) = \frac{Q_i(t)}{Q_s(t)} \tag{XI-39}$$

= fraction of system unavailability contributed by minimal
cut set i   (XI-40)

and:

$$e_k(t) = \frac{\sum_{k \text{ in } i} Q_i(t)}{Q_s(t)} \tag{XI-41}$$

= fraction of system unavailability contributed by
failure of component k.   (XI-42)

The symbol Σ in equation (XI-41) denotes a sum of $Q_i(t)$ over all those minimal cut sets containing component k as one of its components. Because the system can be down if and only if one or more of the cut sets is down, the sum of $Q_i(t)$ in Equation (XI-41) is the probability that the system is down due to component failure k being one of the causes. In terms of conditional probabilities, $E_i(t)$ is approximately the probability that the system is down due to minimal cut set i, given the system is down. The quantity $e_k(t)$ is approximately the probability that the system is down due to component k being one of the causes, given the system is down. (The quantities are approximate because intersections of minimal cut sets are ignored, i.e., the rare event approximation is used.)

When all components of the fault tree are repairable or cyclic, and constant values are used for the component unavailabilities then the importances $E_i(t)$ and $e_k(t)$ are constant and independent of time: $E_i(t) = E_i$ and $e_k(t) = e_k$. The minimal cut set and component importances thus can be ranked from largest to smallest without regard to the time considered.

With regard to the system failure occurrence rate, the minimal cut set importance $\widehat{E}_i(t)$ and the component importance $\widehat{e}_k(t)$ are:

$$\widehat{E}_i(t) = \frac{W_i(t)}{W_s(t)} \tag{XI-43}$$

= fraction of system failure occurrences at
time t contributed by minimal cut set i   (XI-44)

and:

$$\widehat{e}_k(t) = \frac{\displaystyle\sum_{k\ in\ i} W_i(t)}{W_s(t)} \qquad (XI\text{-}45)$$

= Fraction of system failure occurrences at time t
in which component k is one of the contributors.   (XI-46)

The reasoning for the above two formulas is the same as used previously. With regard to $\widehat{e}_k(t)$, component k is defined to be one of the contributors to system failure at time t if it is either down at time t or if it fails at time t. If constant values are used for all the component characteristics then $\widehat{E}_i(t)$ and $\widehat{e}_k(t)$ are again simple constants which can be ranked from highest to lowest without regard to time.

For the convenience of the reader, Tables XI-2 and XI-3 summarize all the formulas which have been presented for the evaluation of a fault tree.

Table XI-2. Summary of Equations for Reliability Characteristics

	Required Data	Unavailability	Occurrence Rate	Unreliability
Components				
λ-model				
Non-repairable	λ	$q(t) = 1 - e^{-\lambda t} \cong \lambda t,\ \lambda t < 0.1$	$w(t) = \lambda e^{-\lambda t} \cong \lambda,\ \lambda t < 0.1$	$F(t) = 1 - e^{-\lambda t} \cong \lambda t,\ \lambda t < 0.1$
Repairable, monitored	$\lambda,\ T_D$	$q(t) = \dfrac{\lambda T_D}{1 + \lambda T_D} \cong \lambda T_D,\ \lambda T_D < 0.1$	$w(t) = \lambda$ (asymptotic)	same as above
Repairable, periodically tested	$\lambda,\ T,\ T_R$	$q(t) = \dfrac{\lambda T}{2} + \lambda T_R \cong \dfrac{\lambda T}{2},\ T_R < 0.1\,T$	$w(t) = \lambda$ (asymptotic)	same as above
p-model				
Cyclic Component	$p,\ n(t),\ k(t)$	$q(t) = 1 - (1 - p)^{n(t)} \cong n(t)p,\ n(t)p < 0.1$	$w(t) = pk(t)$	$F(t) =$ same as $q(t)$
Minimal Cut Sets		$Q_i(t) = \prod\limits_{k=1}^{n_i} q_k(t)$ where n_i = the number of components in the ith minimal cut set	$\begin{aligned} W_i(t) = &\ q_2(t)\,q_3(t) \cdots q_{n_i}(t)\,w_1(t) \\ &+ q_1(t)\,q_3(t) \cdots q_{n_i}(t)\,w_2(t) \\ &+ q_1(t)\,q_2(t) \cdots q_{n_i}(t)\,w_3(t) \\ &\ \cdots \\ &+ q_1(t)\,q_2(t) \cdots q_{n_i-1}(t)\,w_{n_i}(t) \end{aligned}$	$F(t) \cong N_i(t_1, -\ t_2)\ ;\ N_i(t_1,\ t_2) < 0.1$
System		$Q_s(t) = \sum\limits_{i=1}^{N} Q_i(t)$ where N = the number of minimal cut sets	$W_s(t) = \sum\limits_{i=1}^{N} W_i(t)$	$F(t) \cong N_s(t_1,\ t_2);\ N_s(t_1,\ t_2) < 0.1$

where

λ = component failure rate per hour (operating or standby, as applicable)
T_D = average downtime per failure in hours
T = test interval in hours
T_R = average repair time per failure in hours
p = probability of cyclic component failure per demand
$n(t)$ = expected number of demands in time t
$k(t)$ = cyclic component demand rate per hour at time t

Table XI-3. Summary of Equations for Quanitative Importance

	ith Cut Set Importance	kth Component Importance
System Unavailability	$E_i(t) = \dfrac{Q_i(t)}{Q_s(t)}$	$e_k(t) = \dfrac{\sum\limits_{k \text{ in } i} Q_i(t)}{Q_s(t)}$
System Failure Occurrence Rate	$\hat{E}_i(t) = \dfrac{W_i(t)}{W_s(t)}$	$\hat{e}_k(t) = \dfrac{\sum\limits_{k \text{ in } i} W_i(t)}{W_s(t)}$

(h) Sensitivity Evaluations and Uncertainty Analyses

In the previous sections, the computation of point estimates for the unavailability and failure occurrence rate of the top event of a fault tree were described. In this section we briefly take up the question of how to evaluate the sensitivity of these estimates to variations or uncertainty in the component data or models.

Sensitivity studies are performed to assess the impact of variations or changes to the component data or to the fault tree model. It is particularly convenient to assess effects of component data variations using the formulas presented previously in this chapter because they explicitly contain component failure rates, test intervals, and repair times as variables. In sensitivity analyses different values may be assigned these variables to determine the differences in any result. For example, if T is a periodic testing interval, then the effects on system unavailability with regard to different testing intervals can be studied by varying the T's for the components. This can entail as simple a calculation as redoing the computations with different T's or as complex as employing dynamic programming. Likewise, failure rates (λ) can be changed to determine the effects of upgrading or downgrading component reliabilities.

As a type of sensitivity study, scoping-type evaluations can also be performed by using a high failure rate and a low failure rate for a particular event on the tree. If the system unavailability does not change significantly, then the event is not important and no more attention need be paid it. If the system unavailability does change significantly, then more precise data must be obtained or the event must be further developed to more basic causes. A wide spectrum of sensitivity analyses can be performed, depending on the needs of the engineer.

In judging the significance of an effect, it is important that the analyst take into account the precision of his data. For example, although a factor of 2 variation in the system unavailability might be very significant when failure rates are known to 3 significant figures, the same factor of 2 variation would probably not be significant when failure rates are known only to an order of magnitude.

As a type of sensitivity evaluation, formal error analyses can be performed to determine the error spread in any final result due to possible data uncertainties or variabilities. The error spread obtained for the results gives the uncertainty or variability associated with the result. The error analyses employ statistical or probabilistic techniques, which are independent of the fault tree evaluation techniques per se; the discussion therefore, will be short.

A variety of error analysis methods can be used, and we will briefly explain the approach when data are treated as random variables.* For the random variable approach, the method most adaptable to a fault tree evaluation is the Monte Carlo simulation technique. The Monte Carlo method can accommodate general distributions, general sizes of the errors, and dependencies.

In the Monte Carlo method, the fault tree evaluations are repeated a number of times (each repetition is called a "trial") using different data values (e.g., λ's and T_R's) for each calculation. The variation in the data values is "simulated" by randomly sampling from probability distribution functions which describe the variability in the data. The probability distributions can be Bayesian prior distributions on the parameters λ, T_R, etc., or can be distributions representing plant-to-plant variation in the failure rates and other data. Each trial calculation will give one value for the system result of interest such as the system unavailability or occurrence rate. The whole set of repeated calculations will give a set of system results from which an error spread is determined (e.g., picking the 5% largest value and 95% largest value to represent the 90% range for the result).

The above method is completely analogous to repeating an experiment many times to determine the error in the experimental value. The final error spread on the result is the estimate of the result variability arising from the variability in the failure rates and other data treated as random variables. (Section 2b of Chapter XII describes several fault tree-oriented computer codes for performing Monte Carlo simulation.)

---

*When failure rates and other data are treated as constants, with uncertainties arising from the variability of the estimators, then classical confidence bound approaches should be used. The reader is referred to Mann, Schafer, and Singpurwalla (reference [30]) for further details.

CHAPTER XII – FAULT TREE EVALUATION COMPUTER CODES

1. Overview of Available Codes

This chapter covers the computer code methodology currently available for fault tree analysis. The codes discussed are divided into five groups. The numbers in brackets refer to references in the bibliography.

Group 1. Codes for Qualitative Analysis

PREP [42] 1970
ELRAFT [35] 1971
MOCUS [11] 1972
TREEL and MICSUP [29] 1975
ALLCUTS [39] 1975
SETS [46] 1974
FTAP [43] 1978

Group 2. Codes for Quantitative Analysis

KITT1 [42] 1969, KITT2 [40] 1970
SAMPLE [38] 1975, MOCARS [25] 1977, et al.
FRANTIC [41] 1977

Group 3. Direct Evaluation Codes

ARMM [26] 1965
SAFTE [13] 1968
GO [14] 1968, GO "Fault Finder" 1977
NOTED [45] 1971
PATREC [19] 1974, PATREC-MC [20] 1977
BAM [34] 1975, WAM-BAM [22] 1976, WAMCUT [9] 1978

Group 4. A Dual-Purpose Code

PL-MOD [28] 1977

Group 5. Common Cause Failure Analysis Codes

COMCAN [3] 1976
BACKFIRE [5] 1977
SETS [47] 1977

Group 1 consists of codes that perform the qualitative evaluation of a fault tree (i.e., codes that compute minimal cut and/or path sets). The codes in Group 2 perform quantitative (probabilistic) analysis based on the structural information embodied in the cut sets. The codes in Group 3 are designed to perform direct numerical evaluation of a fault tree without computing cut sets as a necessary intermediate step; however, many of them will generate cut sets on request as a nonintegral part of the analysis. PLMOD, a dual purpose code that can be used both for the qualitative and quantitative analysis of a fault tree constitutes Group 4, and, finally, Group 5 contains the codes developed for use in common cause analysis. The five groups of codes are described in the succeeding sections of this chapter.

2.   Computer Codes for Qualitative Analysis of Fault Trees

This section deals with codes that compute the minimal cut (and/or path) sets of a fault tree. The computation of the minimal cut sets is often referred to as the qualitative evaluation of the fault tree because the results are based solely on the structure of the tree and are independent of the probabilities associated with the basic events. In contrast, the probabilistic assessment is called the quantitative evaluation of the fault tree.

The division between qualitative and quantitative aspects develops naturally because the probabilistic analysis often involves repeated evaluation of the tree (i.e., at different time points, using a distribution of failure or repair rates to perform sensitivity or error analysis). Thus, it is often most efficient to perform the time-consuming structural analysis once, save the results in some convenient form (usually minimal cut sets) and then use these results to quantify the tree using different sets of data, as required. Other advantages afforded by the computation of the minimal cut sets are (1) the minimal cut sets themselves provide much useful information to the analyst, even in the absence of any quantitative data, because they indicate the minimal sets of components whose failure will cause the system to fail; (2) non-contributing cut sets (usually based on cut set size) can be discarded prior to quantification, thus increasing computational efficiency and reducing data requirements; (3) the ability to compare the minimal cut sets with the original tree provides a valuable error check; and (4) cut sets are required as part of the input to the common cause analysis codes.

One disadvantage of the minimal cut set codes is that the storage and computer time required to process even medium-size trees can become quite prohibitive. This is because the number of cut sets can increase exponentially with the number of gates and can easily reach the millions or even billions of terms (e.g., one example tree with 299 basic events and 324 gates, had in excess of 64 million cut sets). The problem is complicated further because a simple count of events and gates is a very poor indicator of the expected number of minimal cut sets, and even the number of minimal cut sets may not be a good prediction of required processing time. Thus it is often difficult to predict the storage requirements and run time for a given tree.

Several methods can be used to overcome or at least alleviate the problems associated with obtaining the minimal cut sets. The most common is to eliminate, during the processing, cut sets whose size (number of events) is greater than some prespecified number n. This is often very effective for trees having low order cut sets which dominate the high order cut sets. In WASH-1400 [38] for example, only the single and double event cut sets were retained for the independent failure computations; the higher order cut sets were analyzed only for common mode and common cause failure potentials. Another similar approach is to reduce the tree based directly on cut set probability instead of order. This requires the input of component failure probabilities at the outset, however. Disadvantages of using a tree reduction process are that (1) it is impossible to determine the total failure probability discarded and (2) analysis of dependencies such as events dependent on common causes requires separate evaluation of the higher order cut sets.

Other techniques used in some codes are efficient "packed" and/or bit-level storage schemes, use of auxiliary storage media during cut set processing, and automatic tree decomposition schemes. The latter seems to be a promising method, and is discussed further in later sections (see sections on SETS, FTAP, and PL-MOD).

In the remainder of this section, we discuss the individual qualitative analysis codes. PREP, discussed in section 1(a), was the first cut set code. It is included mainly for background; its algorithm has been superseded by improved methods. ELRAFT, MOCUS, MICSUP, ALLCUTS, SETS, and FTAP discussed in sections 1(b)-(g), all use variations on the "top-down" and "bottom-up" methods described earlier in Chapter VII, section 4. SETS is somewhat different from the other codes in that it provides a very general and flexible tool for manipulating the fault tree in the form of its corresponding Boolean equations.

(a) PREP

The PREP and KITT codes [40] [42], written in FORTRAN IV for the IBM 360 computer and released in 1970, were the first computerized fault tree evaluation codes. PREP is a minimal cut set (or path set) generator, and KITT1 and KITT2 perform time dependent fault tree analysis in the context of Kinetic Tree Theory, using the results from PREP. The KITT codes will be discussed in the quantification section.

PREP consists of two parts: PREP-TREBIL and PREP-MINSET. TREBIL (for "tree build") takes the user's input description of a fault tree and builds a FORTRAN subroutine of the Boolean equations for the tree. MINSET then uses the TREE subroutine produced by TREBIL to find the tree's minimal cut and/or path sets.

PREP-MINSET has two options for cut set generation: COMBO and FATE. COMBO systematically fails all single basic events, pairs of basic events, groups of three basic events, etc., to determine which combinations cause the top event of the fault tree to occur. The user determines the maximal size of the cut sets to be computed (for low probability events, such as those found in nuclear power plant fault trees, doubles or triples usually suffice). FATE incorporates quantitative data on the component's reliability to find minimal cut sets which are "most likely" to occur. It does this by performing Monte Carlo simulation.

The main disadvantage of PREP is that COMBO requires a prohibitive amount of computer time for large order cut sets of large fault trees, whereas FATE is not guaranteed to find all the minimal sets. Also, the input to PREP is limited to AND and OR gates, so NOT gates, both explicit and implicit (e.g., exclusive OR gates), are prohibited; and special gates, such as k-out-of-n gates, must be input in terms of their basic AND and OR gate structure. The basic events are assumed to be independent; unlimited replicated events are allowed; there is no way to generate cut sets for intermediate gates; and there is no easy method to input replicated portions of the tree. PREP allows a maximum of 2000 components and 2000 gates; minimal cut sets found by COMBO are limited to maximum length of 10 components.

(b) ELRAFT

The ELRAFT (Efficient Logic Reduction of Fault Trees) code [35] uses the unique factorization property of the natural numbers to find the minimal cut sets of a fault tree. Every integer greater than 1 can be expressed as a unique (except for order) product of prime factors. In the ELRAFT code, each basic event is assigned a unique prime number. The tree is processed from the bottom up, and cut sets for the gates at successively higher levels are represented by the product of the numbers associated with their input events. The major drawback of ELRAFT is that for large trees, the product of the prime factors can soon exceed the capacity of the machine to represent the number. Coded in FORTRAN IV for the CDC 6600, ELRAFT is capable of finding minimal cut sets of up to six basic events for the top event and other specified intermediate events.

(c) MOCUS

The MOCUS code [11] was written in 1972 to replace PREP as a minimal cut set generator for the KITT codes. The so-called "Boolean indicated cut sets" (BICS) are generated by successive substitution into the gate equations beginning with the top event and working down the tree until all gates have been replaced by basic events.* If the tree contains no replicated events, the BICS will be minimal; otherwise, the nonminimal BICS must be discarded. The MOCUS algorithm may be used to find the minimal cut or path sets for up to 20 gates in a given tree. The user may place an upper limit on the length of cut sets found if desired. Other aspects of the MOCUS code are identical to PREP. MOCUS is written in FORTRAN IV for the IBM 360 series computer.

(d) TREEL AND MICSUP

TREEL and MICSUP [29] are based on an idea similar to that used in MOCUS except that instead of working from the top event down, MICSUP (Minimal Cut Set UPward) starts with the lowest level gate basic inputs and works upward to the top tree event.* TREEL is a preprocessor that checks the tree for errors and determines in advance the maximum number and size of the Boolean indicated cut and path sets. As a result of processing the tree from bottom to top, MICSUP has the advantage of generating the BICS for each intermediate gate of the tree. Nonminimal BICS and BICS of length greater than a user-specified limit can be discarded as they appear, thus reducing the computer time and storage requirements. As with MOCUS, most other aspects of the code are similar to PREP.

(e) ALLCUTS

Another code for finding minimal cut sets is ALLCUTS [39] developed by the Atlantic Richfield Company. ALLCUTS uses a top-down algorithm, similar to that of MOCUS. An auxiliary program BRANCH can be used to check the input and cross reference the gates and input events, and a plot program KILMER can be used to produce a Calcomp plot of the fault tree based on the fault tree input description and conversational plotting instructions. ALLCUTS optionally allows input of basic event probability data. If this data is input, ALLCUTS can compute the top event probability, sort and print up to 1000 minimal cut sets in descending order of probability, and select cut sets in specified probability ranges. ALLCUTS handles up

---

*The basic top-down and bottom-up algorithms for minimal cut set generation are explained in chapter VII, section 4.

to 175 basic events and 425 gate events; the current version of the code uses 110 K, octal. ALLCUTS is written in FORTRAN IV and COMPASS (assembly language) for the CDC 6600 computer.

(f) SETS

The Set Equation Transformation System [46], developed by Sandia Laboratories, is a general program for the manipulation of Boolean equations which can be applied to fault trees and used to find minimal cut or path sets. The advantages of the SETS code are its generality and flexibility, one example of which is the ability to dynamically manipulate the tree via SETS user programs. This capability gives the user a great deal of control over the processing, a feature which can be especially helpful when analyzing large trees. For example, a SETS user program may be written to decompose the original tree and process it in stages without requiring any changes to the original fault tree input description. A recently added feature enables SETS to automatically identify the independent subtrees and select stages for efficient processing of large trees. A packed, bit-level storage scheme and use of auxiliary storage are other SETS features aimed at efficient processing of large trees.

Unlike PREP, ELRAFT, MOCUS, ALLCUTS, and MICSUP, SETS can handle complemented events, exclusive or gates, and special gates represented by any valid Boolean expression defined by the user. It can be used to find the "prime implicants" (a more general term than minimal cut sets which includes the possibility of having both an event and its complement in a Boolean equation) of any intermediate gate. Other useful features are free field input, the ability to easily input replicated subtrees, the option of saving the cut sets or factored equation for any event on a file for future use. The factored equation is a compact form of the minimal cut set equation from which cut sets of any order can be generated for enumeration purposes.

SETS allows tree reduction based on both cut set order and cut set probability. It will also rank and print the minimal cut sets in order of descending probability (assuming basic event probabilities have been input). SETS is written in FORTRAN for the CDC 6600.

(g) FTAP

The Fault Tree Analysis Program [43] is a cut set generation code developed at the University of California Berkeley Operation Research Center. FTAP is unique in offering the user a choice of three processing methods: top-down, bottom-up and the "Nelson" method. The top-down and bottom-up approaches are basically akin to the methods used in MOCUS and MICSUP, respectively. The Nelson method is a prime implicant algorithm which is applied to trees containing complement events and uses a combination of top-down and bottom-up techniques. FTAP is the only fault tree code, other than SETS, which can compute the prime implicants.

FTAP uses two basic techniques to reduce the number of non-minimal cut sets produced and thereby increase the code's efficiency. The first technique, used in the bottom-up and Nelson methods, is modular decomposition. This approach is quite similar to that used in PL-MOD (see section 4) and somewhat similar to the SETS algorithm for identifying and processing independent subtrees (see section 2(f)). The second technique, used in the top-down and Nelson methods, is called the "dual algorithm" in the FTAP documentation [43]. The algorithm involves transformation of a product of sums into a sum of products whose dual is then taken using a special method. The author claims that the non-minimal sets appearing during construction of the dual "will always be less than the number of such sets in [the original product of sums], usually many times less."

Other features of FTAP are the ability to reduce the tree based on cut set order or probability, the ability to find either path sets or cut sets, direct input of symmetric (k-out-of-n) gates, and considerable flexibility and user control over processing and output.

FTAP is written in FORTRAN and assembly language, and is available in versions for the CDC 6600/7600 and IBM 360-370 model computers.

3.   Computer Codes for Quantitative Analysis of Fault Trees

This section deals with codes which perform quantitative evaluations of fault trees. The input to these codes consists of two parts:

(1) the equation for the top event unavailability or unreliability (usually from the minimal cut sets but can be obtained from other non-fault tree models such as block diagrams or schematics)

(2) failure rate, test, and repair data for the components appearing in the equation.

Given the above inputs, several types of quantitative results may be computed including:

Numerical probabilities: probabilities of system and component failures;
Quantitative importances: quantitative rankings of contributions to system failure;
Sensitivity evaluations: effects of changes in models and data, error bounding;

The KITT and FRANTIC codes, described in sections 2(a) and 2(c), respectively, compute time-averaged and time-dependent point estimates for the system failure probability. KITT also computes quantitative importances. SAMPLE and MOCARS, described in section 2(b), compute a distribution and error bounds for system failure probability based on uncertainty, error, or variation in the component failure characteristics.

(a) The KITT codes

KITT1 and KITT2 [40] [42] perform time dependent fault tree quantification based on the minimal cut or path set description of the tree. The codes can thus be used in conjunction with any qualitative analysis code which generates the minimal cut sets in terms of components (basic events) such as PREP, MOCUS, SETS, etc. PREP and MOCUS generate the cut sets in a form which is directly usable as input to the KITT codes. Other required inputs are the component failure rates and repair characteristics. Components are assumed to have exponential failure distributions. Each component may have a constant repair time, an exponential repair distribution, or may be nonrepairable. In addition, KITT2 allows each component to have its own unique time phases whereby its failure and repair data may vary from phase to phase.

The KITT codes compute the following five probability characteristics for the system failure (top event), each component, and each minimal cut or path set at arbitrary time points specified by the user:

The probability of the failure existing at time t (the unavailability).
The probability of the failure not occurring to time t (the reliability).
The expected number of failures occurring to time t.
The failure rate per hour.
The occurrence rate per hour.

In addition to the above, the KITT codes rank the events in single and double component cut sets by qualitative and quantitative importance. See Chapter XI for a discussion of importance measurements.

(b) SAMPLE, MOCARS, et al.

Several codes have been written to compute the probability distribution of a calculated system result (such as unavailability) when probability distributions are assigned to the component failure rates to account for data variability. These codes use Monte Carlo simulation in which the component failure rates are sampled from input probability distributions. The sample values for the failure rates are then combined by means of a system function given in a user-supplied FORTRAN subroutine to determine the sample system results. After a number of these "trials," the different system values can be tabulated and the resulting empirical distribution can be characterized. By this method, the effect on the system unavailability of uncertainties or variations in the component failure rates can be assessed.

Typically, the user-supplied system unavailability function might be the minimal cut set equation obtained from one of the qualitative analysis codes.

SAMPLE [38] was the Monte Carlo code used in WASH-1400. SAMPLE allows normal, log-normal, or log-uniform distributions to be specified for the component failure rates. The output distribution is presented in terms of estimated empirical probability percentiles from which the estimated median and upper and lower bounds can easily be read. The output also includes the estimated mean and standard deviation of the distribution and a tabular histogram of the system density function. Sample is written in FORTRAN IV.

MOCARS [25] is similar to principle and operation to SAMPLE, but allows a larger variety of sampling distributions including exponential, normal, gamma, beta, lognormal, binomial, Poisson, Weibull, and empirical distributions. It allows the system unavailability function to be specified either as FORTRAN statements or in terms of cut sets. Other options include microfilm plotting using the Integrated Graphics System (IGS) and the ability to perform a Kolmogorov-Smirnov goodness-of-fit test on the output distribution to see if it resembles a normal, lognormal, or exponential function. MOCARS was written in FORTRAN to run on the INEL CDC 76-1973 operating system.

Other expanded versions of SAMPLE [4] have been written, but they are all quite similar, and so we will not discuss them here.

(c) FRANTIC

The FRANTIC (Formal Reliability Analysis including Normal Testing, Inspection and Checking) code [41] computes the average and time dependent unavailability of any general system model such as a fault tree or event tree, incorporating in detail the effects of different periodic testing schemes. The program can be used to assess the effects on system unavailability of test downtimes, repair times, test efficiency, test bypass capabilities, test-caused failures, and different test staggerings. In addition to periodically tested components, nonrepairable and monitored components as well as human error and common cause contributions can also be modeled.

As in the SAMPLE code, the system model function is input in the form of FORTRAN subroutine. For each component, the failure rate, and test and repair characteristics must be provided. Exponential failure distributions are assumed. Other input includes the time period for the calculations, and print and plot options. Calcomp plots of the time dependent system unavailability function may be produced.

A Monte Carlo version of the FRANTIC code [16] is available in which sampling distributions may be input for the component failure rates. FRANTIC is written in FORTRAN IV for the IBM 360-370 series computers.

4.   Direct Evaluation Codes

As their name implies, the direct evaluation codes quantify the system model in a single step. Thus they do not produce cut sets as an integral part of the analysis and they require probabilistic input for each component from the outset of the processing. The output from these codes is generally in the form of point estimates for the system unavailability or failure probability.

The GO and WAM-BAM codes, described in sections 3(c) and 3(f) respectively, offer the advantage of allowing complement events and some modeling of dependencies. GO also allows switches and time delays and models all system states instead of a single fault event. Both GO and WAM-BAM reduce storage requirements by eliminating low probability paths at an intermediate stage of the processing and at the same time keep track of the total of the discarded path probabilities. Disadvantages are mainly connected with the inability to produce cut sets (many of the direct evaluation codes have added the option to compute cut sets, but not as an integral part of the analysis), and the necessity of inputting probabilities for all of the components even though many may be insignificant contributors to system failure. Also, a change in probabilities often requires a complete rerun.

(a) ARMM

The ARMM (Automatic Reliability Mathematical Model) code [26], developed by North American Aviation for the U.S. Air Force and modified by Holmes and Narver for application to nuclear power plant systems, was the first direct evaluation code. ARMM models a reliability block diagram using a success path approach. The component failure probabilities are determined using failure density functions supplied for each component. The program is capable of handling Weibull (time-dependent failure rate) density functions, dependent components, and mutually exclusive failure modes. It is written in FORTRAN IV for the IBM 360 computer.

(b) SAFTE

The SAFTE (Systems Analysis by Fault Tree Evaluation) codes [13], SAFTE1, SAFTE2, and SAFTE3, are Monte Carlo simulation programs using techniques similar to the FATE option of PREP to generate random times to failure of components in a fault tree. However, instead of computing the cut sets, the SAFTE codes directly generate a distribution of times to failure for the system. The SAFTE1 code works as described above; the SAFTE2 code also includes the ability to sample from normal repair distributions to get a time to repair for each component. In this version, a failed component may be repaired (made as good as new) before system failure, and then resume operation with a new random time to failure and time to repair. In both SAFTE1 and SAFTE2, the random times to failure are generated from exponential failure distributions.

SAFTE3 computes the probability of system failure based on steady state repair, using either direct or importance sampling techniques. The SAFTE codes are written in FORTRAN IV for the IBM 360 computer.

(c) GO

The GO methodology [14], developed in the mid-1960's by Kaman Sciences Corporation, differs from the fault tree approach in that the normal operating sequence is modeled and all possible system states are considered. The input model used is called a GO chart and it resembles a schematic or flowchart made up of a set of standardized operators which describe the logical operation and interconnection of the system components. Some of the 16 GO operators are similar to fault tree gates, but in addition to logic functions, time delays and switches can be modeled as well as complementary event logic and mutually exclusive states. GO also provides a simplified method for modeling repeated portions of the tree through the use of "supertypes." In addition to specifying the types of operators and their interconnections, the user also specifies the probabilities associated with the possible operational modes of each component. This process is analogous to supplying failure probabilities for components in a fault tree; however, in the GO approach, probabilities are given for states other than simple success or failure (e.g., the probability of premature operation, or the probabilities of response over a series of time points are supplied for some operators). The outputs from GO are the probabilities of occurrence of individual output events or the joint probability density of degrees of performance for several output events. The output events can include system success, and various degrees of failure such as spurious or premature operation, delayed or partial operation, and complete failure to operate. The effects of component repair cannot be modeled.

The numerical evaluations are performed in a one-step process as the signals (event probabilities) are traced through the model using a Markov chain (event tree) approach to propagate the values. This means that a change in component probabilities, such as for sensitivity studies, requires a complete reevaluation even though the system structure remains unchanged. Because the probability tree can easily become very large, GO has options for pruning the tree of branches with probabilities lower than a selected value, and of deleting signals which are no longer needed, while keeping track of the total probability of the discarded paths. GO also includes a "Fault Finder" option to compute the 4th order cut sets for a selected output event.

Because of the diversity and detail of the GO operators, and the necessity of including all system components, the modeling process for GO is somewhat more complex than that for fault trees. However, it may be argued that because GO charts are similar to the familar system schematics, the modeling process is easily learned by designers and engineers. If the analyst wishes to use the fault tree model instead of the GO chart, it is still possible to evaluate the tree using GO. In this case only the subset of the GO operators analogous to the fault tree gates would be used and the output would be a point estimate of the top event failure probability.

GO is written in FORTRAN for the CDC 7600.

(d) NOTED

NOTED [45], developed by the United Kingdom Atomic Energy Authority in 1971, is similar in concept to GO. However, instead of analyzing the system at a series of discrete time points, NOTED produces a graph of the cumulative failure probability as a continuous function of time at any of several points in the system. Similarly, the behavior of the input components is described by continuous failure distributions including exponential lognormal, normal, Weibull and forms including repair times.

(e) PATREC

PATREC [19] was the first computer code to apply list processing techniques to fault tree evaluation. Rather than generating cut sets, PATREC evaluates the tree directly, using a pattern recognition algorithm implemented in the PL/1 programming language. A set of subtree patterns along with their corresponding probability equations are stored in the computer code's library. The fault tree is then searched for occurrences of the library patterns. Each recognized pattern is replaced by a supercomponent with probability of occurrence equal to that stored in the library. The whole tree is thus eventually reduced to a single leaf which corresponds to the probability of failure of the total system.

PATREC can evaluate trees containing both an event and its complement; direct input of k-out-of-n gates is also supported. Cut sets can be generated, if desired, using an algorithm similar to that of MOCUS, but they are not used in the evaluation of the fault tree. PATREC's greatest limitation is in its handling of replicated events. The pattern recognition scheme yields the correct probabilities only when no events are replicated. Therefore, PATREC replaces a single fault tree having r different replicated events by 2^r fault trees with no replicated events. Even with approximations which allow some of the 2^r fault trees to be discarded, PATREC cannot efficiently evaluate fault trees with more than about 20 replicated events.

PATREC is capable of performing time dependent system unavailability analysis where each basic event may have a failure distribution which is exponential, Weibull, normal, or log normal. For the exponential failure case, the component can optionally be assumed repairable with an exponential repair distribution. In addition, the user can include a constant "failure-on-demand" (e.g., failure to start) probability which is independent of time.

PATREC-MC [20] is a Monte Carlo version of PATREC which can be used to assess the effects of uncertainty in the component reliability parameters. The operation of the code is similar to that of SAMPLE (see section 2(b) of this chapter) except for the representation of the system function. In PATREC-MC, a calculation is first made to identify the patterns in the tree using the list processing methods described earlier. The patterns are then stored in memory so that they can be repeatedly evaluated during the Monte Carlo trials. Note that storage of the patterns for subsequent reevaluation means that PATREC-MC is not really a direct evaluation code because the stored patterns are actually the result of an intermediate qualitative analysis independent of the component probabilities. This distinction will come up again in our discussion of PL-MOD (see section 4 of this chapter).

(f) WAM-BAM

The WAM-BAM codes [9] [22] [34] were developed at Science Applications Inc. for EPRI beginning in 1975. The WAM-BAM package actually consists of four codes: WAM, WAMTAP, BAM, and WAM-CUT. WAM and WAMTAP are input preprocessors for the evaluation code BAM (Boolean Arithmetic Model). The WAM preprocessor, like PREP, is designed to ease the input preparation process. It generates the numeric input for BAM from the input description of the fault tree and the event probabilities. At the user's option, the input to BAM can be saved and subsequently modified by WAMTAP. WAMTAP allows the probability of single components or groups of components to be changed in order to run sensitivity studies or to include common cause contributions. WAM-CUT can be used to compute minimal cut sets, and the mean and variance of the probability of any gate. It can also generate the input to a Monte Carlo code, SPASM, which computes a distribution on gate probability.

The evaluation code, BAM, uses a combination of concepts from the GO methodology and fault tree analysis. The GO computational scheme is used but the operations are modeled as gates on a fault tree. Eight possible logical combinations of 2 events and their complements are included as allowable gates. In BAM, the probability of the top event is computed by forming a truth table, each line of which represents a product term (P term) event disjoint from all the other P terms. In terms of the GO methology, the P terms are equivalent to paths in the GO event tree. The output from BAM is a point probability of the top event. As mentioned earlier, WAMTAP can be used to modity the input for BAM sensitivity studies. WAM-BAM is written in FORTRAN for the CDC 6600.

5.   PL-MOD: A Dual Purpose Code

PL-MOD [28] is described separately in this section because, though it can perform both a qualitative and quantitative fault tree analysis, it depends neither on the standard cut set generation nor the direct evaluation techniques. Like the PATREC-MC code, it performs a qualitative analysis which does not rely on standard cut set generation techniques, but which can be used repeatedly for quantification.

The PL-MOD computer code works by "modularizing" the fault tree directly from a description of its component and gate diagram. Defined in terms of a reliability network diagram, a module is a group of components which behaves as a supercomponent (i.e., it is completely sufficient to know the state of the supercomponent, and not the state of the components which comprise it, to determine the state of the system). In terms of a fault tree diagram, an intermediate gate is a module to the top event tree if none of the basic events contained in the gate domain (i.e., all branches below the gate) appear elsewhere in the fault tree. Briefly, modularization implies that all the independent subtrees (i.e., modules or subsystems) are identified, and the minimal cut sets are defined recursively in terms of these modules. Or, to put this is a slightly different way, a modularized tree is one which is equivalent to the original tree but in some sense "maximizes" the decomposition of the tree into independent subtrees.

The concept and advantages of modularization have been known for some time [2] and an algorithm for finding the finest modular decomposition of a fault tree given its cut sets was described by Chatterjee [6] in 1975. The modularization process used by PL-MOD is, as mentioned earlier, unique in that it is applied not to the cut sets, but directly to a description of the fault tree diagram, using the list processing features of the PL/1 programming language. The modularization process used by PL-MOD is somewhat complex, and will therefore not be discussed here. (A complete description appears in reference [28]).

Features of PL-MOD are the ability to handle complemented events, direct input of symmetric (k-out-of-n) gates, free field input, and dynamic storage allocation. The output from PL-MOD includes the standard and modular minimal cut sets for the top event and specified intermediate gates of the tree.

Some disadvantages of PL-MOD are its machine dependence (PL/1 is not available in many computer systems) and the lack of familiarity with PL/1 among scientific users.

The quantitative capabilities of PL-MOD include the computation of the occurrence probability and importance (see Chapter XI) for the top event and all other modules. PL-MOD also has a Monte Carlo option for computing uncertainties and time-dependent unavailability evaluation option which can handle non-repairable, repairable (revealed fault), and periodically tested components.

6.   Common Cause Failure Analysis Codes

Common cause failure analysis is becoming increasingly important in system reliability and safety studies because it has been recognized that common cause failures can often dominate the random hardware failures. Common cause failure analysis attempts to identify the modes of system failure (i.e., minimal cut sets) which have the potential of being triggered by a single, more basic common cause; the minimal cut sets which need to be identified are those with two or more events, all of which are susceptible to a single common cause failure mechanism.

(a) COMCAN

COMCAN [3], developed at INEL, was the first program to perform common cause failure analysis. The input to the program consists of two parts: (1) the minimal cut sets of the fault tree to be analyzed, and (2) the common cause susceptibility data for each basic event. The output from the program is a list of the minimal cut sets which are common cause candidates.

A minimal cut set may be identified as a common cause candidate by either of two criteria. The first criterion requires that all the events in the cut set be potentially affected by the same cause or condition. The second criterion requires that all the events in the cut set share susceptibility to a common cause or condition and in addition, all components implied by the basic events in the minimal cut set must share a common physical location with respect to the common cause susceptibility. Some typical common causes include: impact, vibration, pressure, grit, stress, and temperature. The minimal cut sets and common cause susceptibility data constitute the required inputs. Optional inputs are the component manufacturers, location domain definitions for generic causes, the location of each component implied by the basic events, and ranks of component susceptibility to each common cause. The more input provided, the more refined will be the search for the common cause candidates. The output options include the ability to print only the common cause candidates with ranks $\geqslant N$, and to include similar type components as one of criteria for common cause candidates.

COMCAN is written in FORTRAN IV for the IBM 360 computer.

(b) BACKFIRE

The BACKFIRE code [5], published in May 1977, is an offshoot of COMCAN. The required and optional inputs are almost the same except that BACKFIRE permits more than one location to be specified for a component. This is useful for piping and wiring which may pass through domain barriers. Like COMCAN, BACKFIRE is written in FORTRAN IV for the IBM 360 computer.

(c) SETS

The SETS code, described in section 1(f), can also be used for common cause analysis [47]. The analysis is conducted in a manner similar to that of COMCAN, by inputting generic cause susceptibilities for each basic event. A variable transformation incorporates the common cause susceptibilities into the Boolean equation for the top or any intermediate gate of the fault tree, and a few simple manipulations allow the user to display the cut sets which are the common cause candidates.

BIBLIOGRAPHY

1. Army Ordnance Corps, *Tables of the Cumulative Binomial Probabilities*, ORDP 20-11, 1952.

2. Z. Birnbaum, "On the Importance of Different Components in a Multicomponent System in Multivariate Analysis-II," P.R. Krishnaiah, Editor, Academic Press, New York, 1969.

3. G.R. Burdick, N.H. Marshall and J.R. Wilson, "COMCAN–A Computer Program for Common Cause Analysis," Aerojet Nuclear Company, ANCR-1314, May 1976.

4. J.J. Cairns and K.N. Fleming, "STADIC: A Computer Code for Combining Probability Distributions," General Atomic Co., GA-A14055, March 1977.

5. C.L. Cate and J.B. Fussell, "BACKFIRE–A Computer Code for Common Cause Failure Analysis," University of Tennessee, Knoxville, Tennessee, May 1977.

6. P. Chatterjee, "Modularization of Fault Trees: A Method to Reduce the Cost of Analysis," *Reliability and Fault Tree Analysis*, SIAM, 1975, pp. 101-126.

7. W.C. Cochran, *Sampling Techniques*, John Wiley and Sons, Inc., New York, 1977.

8. W.J. Conover, *Practical Nonparametric Statistics*, John Wiley and Sons, Inc., New York, 1971.

9. R.C. Erdmann, F.L. Leverenz, and H. Kirch, "WAMCUT, A Computer Code for Fault Tree Evaluation," Science Applications, Inc., EPRI NP-803, June 1978.

10. W. Feller, *An Introduction to Probability Theory and Its Applications*, Vol. I, John Wiley and Sons, Inc., New York, 1957, 1968.

11. J.B. Fussell and W.E. Vesely, "A New Methodology for Obtaining Cut Sets for Fault Trees," *Trans, ANS*, Vol. 15, p. 262, 1972.

12. J.B. Fussell and G.R. Burdick, ed., *Proceedings of the International Conference on Nuclear Systems Reliability and Risk Assessment*, Gatlinburg, Tennessee, SIAM, 1975.

13. B.J. Garrick, "Principles of Unified Systems Safety Analysis," *Nucl. Engr. and Design*, Vol. 13, No. 2, pp. 245-321, August 1970.

14. W.Y. Gateley, D.W. Stoddard and R.L. Williams, "GO, A Computer Program for the Reliability Analysis of Complex Systems," Kaman Sciences Corporation, Colorado Springs, Colorado, KN-67-704(R), April 1968.

15. B.V. Gnedenko, Y.K. Belyayev, and A.D. Solovyev, *Mathematical Methods of Reliability Theory*, Translation edited by R.E. Barlow, Academic Press, New York 1969.

16. F.F. Goldberg and W.E. Vesely, "Time Dependent Unavailability Analysis of Nuclear Safety Systems," presented at the National Conference on Reliability, Nottingham, England, September 1977.

17. A.E. Green and A.J. Bourne, *Reliability Technology*, Wiley-Interscience, London, 1972.

18. IEEE, *Guide for General Principles of Reliability Analysis of Nuclear Power Generating Protection System*, IEEE Std. 352-1975, 1975.

19. B.V. Koen and A. Carnino, "Reliability Calculations with a List Processing Technique," *IEEE Transactions on Reliability*, Vol. B-23, No. 1, April 1974.

20. B.V. Koen, A. Carnino, et al., "The State of the Art of PATREC: A Computer Code for the Evaluation of Reliability and Availability of Complex Systems," presented at the National Reliability Conference, Nottingham, England, September 1977.

21. H.E. Lambert, "Fault Trees for Decision Making in Systems Analysis," Lawrence Livermore Laboratories, DCRL-51829, 1975.

22. F.L. Leverenz and H. Kirch, "User's Guide for the WAM-BAM Computer Code," Science Applications, Inc., EPRI 217-2-5, January 1976.

23. D.K. Lloyd, M. Lipow, *Reliability: Management, Methods, and Mathematics*, Prentice-Hall, Inc., Englewood Cliffs, New Jersey, 1962.

24. N.R. Mann, R.E. Schafer and N.D. Singpurwalla, *Mathematical Methods for Statistical Analysis of Reliability and Life Data*, John Wiley and Sons, Inc., New York, 1974.

25. S.D. Matthews, "MOCARS: A Monte Carlo Simulation Code for Determining the Distribution and Simulation Limits," TREE-1138, July 1977.

26. C.W. McKnight, et al., "Automatic Reliability Mathematical Model," North American Aviation, Inc., Downey, California, NA 66-838, 1966.

27. National Bureau of Standards, *Tables of the Binomial Probability Distribution*, App. Math. Series, Vol. 6, 1949.

28. J. Olmos and L. Wolf, "A Modular Approach to Fault Tree and Reliability Analysis," Department of Nuclear Engineering, Massachusetts Institute of Technology, MITNE-209, August 1977.

29. P.K. Pande, et al., "Computerized Fault Tree Analysis: TREEL and MICSUP," Operational Research Center, University of California, Berkeley, ORC 75-3, April 1975.

30. C.R. Rao, *Linear Statistical Inference and Its Applications*, John Wiley and Sons, Inc., New York, 1977.

31. N.H. Roberts, *Mathematical Methods in Reliability Engineering*, McGraw-Hill, New York, 1964.

32. H.G. Romig, *Binomial Tables*, John Wiley and Sons, Inc., New York, 1953.

33. A. Rosenthal, "A Computer Scientist Looks at Reliability Computations," SIAM, 1975, pp. 133-151.

34. E.T. Rumble, F.L. Leverenz, Jr. and R.C. Erdmann, "Generalized Fault Tree Analysis for Reactor Safety," Electric Power Research Institute, Palo Alto, California, EPRI 217-2-2, Interim Report, June 1975.

35. S.N. Semanderes, "ELRAFT, A Computer Program for the Efficient Logic Reduction Analysis of Fault Trees," *IEEE Trans. on Nucl. Sci.*, Vol. NS-18, No. 1, pp. 481-487, February 1971.

36. M.L. Shooman, *Probabilistic Reliability: An Engineering Approach*, McGraw-Hill, New York, 1968.

37. C.P. Tsokos and I.N. Shimi, ed., *The Theory and Applications of Reliability with Emphasis on Bayesian and Nonparametric Methods*, Academic Press, New York, 1977.

38. U.S. Nuclear Regulatory Commission, "Reactor Safety Study—An Assessment of Accident Risks in U.S. Commercial Nuclear Power Plants," WASH-1400 (NUREG-75/014), October 1975.

39. W.J. Van Slyke and D.E. Griffing, "ALLCUTS, A Fast, Comprehensive Fault Tree Analysis Code," Atlantic Richfield Hanford Company, Richland, Washington, ARH-ST-112, July 1975.

40. W.E. Vesely, "Analysis of Fault Trees by Kinetic Tree Theory," Idaho Nuclear Corporation, Idaho Falls, Idaho, IN-1330, October 1969.

41. W.E. Vesely and F.F. Goldberg, "FRANTIC — A Computer Code for Time Dependent Unavailability Analysis," U.S. Nuclear Regulatory Commission, NUREG-0193, October 1977.

42. W.E. Vesely and R.E. Narum, "PREP and KITT Computer Codes for the Automatic Evaluation of a Fault Tree," Idaho Nuclear Corporation, Idaho Falls, Idaho, IN-1349, 1970.

43. R.R. Willie, "Computer Aided Fault Tree Analysis," Operations Research Center, University of California, Berkeley, August 1978.

44. R.L. Wine, *Statistics for Scientists and Engineers*, Prentice-Hall, Englewood Cliffs, New Jersey, 1964.

45. E.R. Woodcock, "The Calculation of Reliability of Systems: The Program NOTED," UKAEA Authority, Health and Safety Branch, Risley, Warrington, Lancashire, England, AHSB(S) R 153, 1971.

46. R.B. Worrell, "A SETS Users' Manual for the Fault Tree Analyst," Sandia Laboratories, Alburquerque, New Mexico, NUREG/CR-0465, SAND 77-2051, 1978.

47. R.B. Worrell and D.W. Stack, "Common Cause Analysis Using SETS," Sandia Laboratories, Alburquerque, New Mexico, SAND 77-1832, 1977.

☆U.S. GOVERNMENT PRINTING OFFICE: 1992-318-277/60129

www.ingramcontent.com/pod-product-compliance
Lightning Source LLC
Chambersburg PA
CBHW061814210326
41599CB00034B/6996